BILL NYE THE SCIENCE GUY'S®
CONSIDER
THE FOLLOWING

*A Way Cool Set of Science Questions,
Answers, and Ideas to Ponder*

Printed in the United States of America

First Edition

1 3 5 7 9 10 8 6 4 2

Library of Congress Catalog Card Number: 95-68629

ISBN: 0-7868-4054-4 (pbk.)
ISBN: 0-7868-5035-3 (lib. bdg.)

Book design by Matthew Van Fleet.

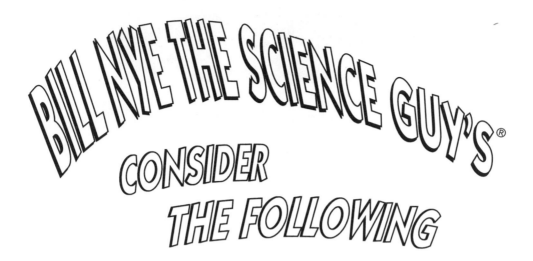

BILL NYE THE SCIENCE GUY'S®

CONSIDER

THE FOLLOWING

*A Way Cool Set of Science Questions,
Answers, and Ideas to Ponder*

by Bill Nye

with additional writing by
Ian Saunders

and illustrations by
Anton Kimball

DISNEY PRESS

New York

ACKNOWLEDGMENTS

The thing that sets us humans apart from other animals is that we can store information outside of our bodies. We share information with other members of our species. We have to. Otherwise humans wouldn't be where they are today. We wouldn't be around.

Most of us learn a lot about the world from our teachers. I've had some great ones. I'd like to thank them all. In this book, I'd particularly like to thank Mr. Lawrence and Mr. Morse, who got me excited about math; Ms. Hrushka, my chemistry teacher; and Mr. Cross, my biology teacher. They gave me plenty to think about.

This book would never have come together without my good friend and colleague Ian Saunders, who helped with polishing my made-for-TV thought processes. Especially, I'd like to thank my family for continuously showing me how important learning and science are. Thinking about the world makes it so that we can live our lives. It's fun to think about and to consider science.

Ian Saunders gratefully acknowledges the help he received while working on this book: Linley Storm, Michael Woodruff, Ted Ridgway, Quigley Raleigh, Jason Gonsky, Erren Gottlieb, Leanne Bunas, James McKenna, Heidi Sowards, Eve Enslow, Alissa Kozuh, Jason Hornick, and Hamilton McCullogh all proved to be indispensible. Like Bill, he would also like to thank his teachers, especially Harry Canning, Tom Mullen, Douglas Saunders, Donald Ferguson, Frederic Saunders, Andrea Rushing, Dale Jackson, Stephen George, Sandra Grant, and John Wilkes, who, like all great teachers, continue to inspire even now. One million thank-yous are due Thea Feldman, the most supportive editor anyone could ask for. Finally, merci beaucoup to the man behind this book, William S. Nye, who really is the coolest Science Guy ever.

TABLE OF CONTENTS

CHAPTER — PAGE — CHAPTER — PAGE

INTRODUCTION

Science is the way that humans figure things out. There isn't actually any one way that scientists always go about it. We learn things with some scientific guidelines and methods, some plans and experiments, and quite a bit of just plain luck. But there is one part of discovery that is always there: we think. We think about what we see, what we hear, what happened before, and what's different from what we might have at first expected.

This collection is based on segments I wrote for my television show, *Bill Nye the Science Guy*. (That's me.) Each show has one topic. We talk about only one big idea in science in each show. For me, though, with every idea there is at least one thing that I just plain wonder about, some big question that I always come back to. So please look over these articles and experiments and think about them. Please consider the following. . . .

Flight

Flying High with Bill Nye

What keeps a heavy airplane from falling to the ground?

PLEASE CONSIDER THE FOLLOWING:

Airplanes are held up by air pressure. Air is invisible, and it's thin. But it can be pretty strong. You can feel the pressure of air on your face when you walk into the wind or ride your bike. And air has pressure because it is made of something called *molecules*.

You've probably heard of molecules. They are the really tiny pieces of stuff that *everything* is made of. You, me, chocolate bars, and air are all made of molecules. Molecules are always bouncing around, like kernels in a popcorn popper. They are always moving

and vibrating. When air molecules bump into things, they make air pressure.

Airplanes stay up in the air because of air pressure and a little something called *Bernoulli's* (burn-OOOH-leez) *principle*.

Bernoulli was a scientist who first realized that fast-moving air creates high air pressure in the direction the air is moving. But as air molecules move faster, they exert less pressure to the sides. You can feel this by blowing on the palm of one hand and bringing the index finger of your other hand close to your mouth. You feel strong air pressure in your palm, but almost nothing on the finger near your mouth.

Air exerts more pressure on things it runs into head-on—like your face, when you're riding a bike. And fast-moving air molecules exert less pressure on things they stream past—like the sides of your head.

Airplanes stay up in the air because of the shape of their wings and the angle that the air hits them. They are usually curved on top and sometimes they curve

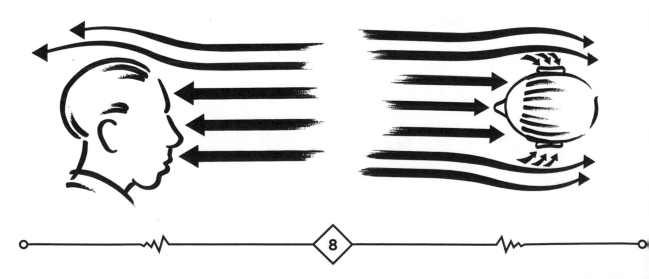

in underneath. The wing makes air molecules go slower under the wing than they do when they go over the wing. When the air molecules are going slowly, they bump up against the wing more, and push harder.

Bernoulli's principle is a way of saying that the faster air molecules moving over the top of an airplane wing push hardest in the direction that they're going. So they don't push down as hard as the slow-er air molecules going past the underside of the wing, which push up. So under the wings, there's higher pressure. That's how planes get "lifted" up. Pretty cool.

This idea of different pressures on opposite sides of things is what makes wings lift and baseballs curve. With big enough wings, a 40-ton 747 jet can stay up in the air, just by air pressure.

On the next page is an uplifting experiment you can try that shows air pressure in action.

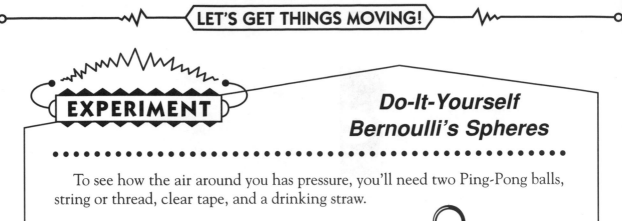

EXPERIMENT

Do-It-Yourself Bernoulli's Spheres

To see how the air around you has pressure, you'll need two Ping-Pong balls, string or thread, clear tape, and a drinking straw.

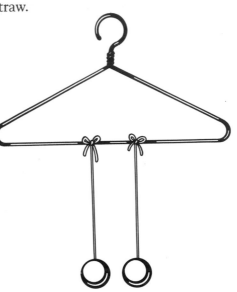

◆ Tape a piece of string to each Ping-Pong ball, using as little tape as possible.
◆ Make sure that the pieces of string are exactly the same length.
◆ Tie the other end of each string to a clothes hanger or the bar you hang your clothes on in the closet.
◆ The balls should hang next to each other, about an inch (2.5 centimeters) apart.
◆ Now use the straw to blow air between the balls.

What do you think will happen?
You might think that the air going between the balls would push them apart. But when you try it, the Ping-Pong balls get closer. In fact, they slap together! Why? Because *fast*-moving air has a *lower* air pressure on things it flows past than *slow*-moving air does. When you blow air through the straw, you make air move *fast*. The air pressure between the Ping-Pong balls goes way down. And since the air on the *outside* of the balls isn't moving very much at all, it has a *higher* air pressure, and it pushes the balls *together*.

This effect is called the Bernoulli principle, named after the Swiss guy who discovered it—not that you have to be Swiss to be a scientist. The Bernoulli principle explains why birds, airplanes, and kites fly. And just so you know, Switzerland is also famous for wristwatches, rich chocolate, cuckoo clocks, and cheese.

Pressure

Vacuums Are All About Nothing

Air pressure acts in all directions. But what if there's no air at all?

PLEASE CONSIDER THE FOLLOWING:

When there's a place or a space with no air, we call it a *vacuum*. *Vacuum* comes from an old word that means "empty." Usually, things that we call empty aren't really *empty*. They're filled with air. An empty room may not have any books or chairs or goldfish in it, but it's got wall-to-wall air! And you know that air is made up of molecules and that these molecules go around bumping into things, making pressure. A vacuum, though, is *really* empty. I mean empty of everything—*including* air molecules. And when there are no air molecules, we get no air pressure.

A good example of a vacuum is a suction cup stuck to a wall. When you push down on the suction cup to attach it to the wall, you force almost all of the air out from under it. That creates a vacuum between the suction cup and the wall. At the same time, the air pressure in the room pushes against the other side of the suction cup. The vacuum under the suction cup has no air molecules and no air pressure, so it doesn't push back. The air molecules in the room hold the cup tightly to the wall.

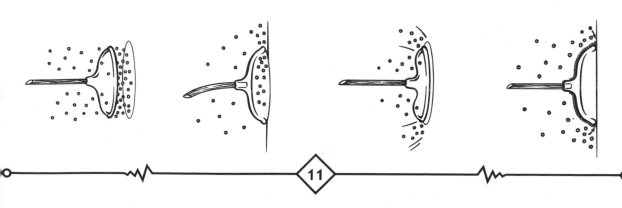

The vacuum under a suction cup doesn't last very long. Air molecules in the room are constantly pushing against all sides of the suction cup. Since there's a vacuum under the suction cup, there's nothing to push back and keep air molecules from getting through tiny gaps between the wall and the cup. That means that some of the air molecules from the room can slowly sneak through into the space under the suction cup. Eventually enough molecules will sneak in, and the suction cup will fall off.

That's why we sometimes wet a suction cup to make it stick better. The liquid fills in tiny gaps between the rubber and the wall. They're small to us, but they're plenty big enough for air molecules to slide through. See, it's harder for air to get through the gaps when they're filled with water, so the vacuum lasts longer.

Soda straws are another way to see a vacuum in action. When you suck up a milkshake, you suck most of the air molecules from the straw into your mouth, making a vacuum in the straw. Then the air pressure in the room pushes down on the top of the milkshake and pushes it up through your straw. Right into your mouth. *Mmmm!*

Of course, when people talk about a vacuum, they usually mean a machine that cleans the floor. But that kind of vacuum works because it has a vacuum inside of it. (Coincidence? You decide.) See, the motor inside the vacuum cleaner blows almost all of the air *out* of the machine, making a partial vacuum *inside* the machine. Then, air in the room rushes in to fill the empty space the only way it can— through the end of the long hose you move over the carpet. All that air rushing in carries dirt up off the floor. *Va-va-va-voom!*

Now, I don't want to put any pressure on you, but there's an experiment on the next page that you can try if you want to see air pressure in action.

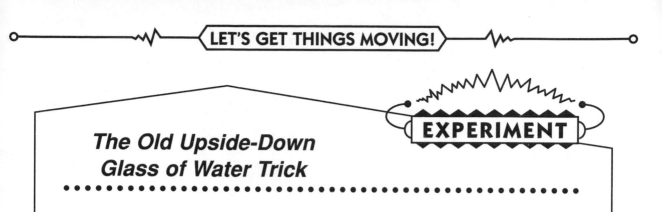

EXPERIMENT

The Old Upside-Down Glass of Water Trick

Which is stronger, water or air? This experiment shows just how strong air pressure can be.

◆ Fill a clear plastic cup with water.

◆ Take a postcard, wet it slightly, and put it on the top of the cup.

◆ Holding the postcard against the cup, turn the whole thing upside down.

◆ Carefully remove your hand from the postcard.

You'll see a space inside the cup—between the bottom of the cup and the top of the water. That space is a vacuum—there's no air there, and no air pressure. But there's lots of air in the room, outside the cup. The pressure of the air in the room pushes *up* against the postcard so it sticks to the cup and holds the water inside.

Hey, it's not magic; it's science.

Buoyancy

Buoy, oh, Buoyancy

Why doesn't a heavy metal ship sink to the bottom of the ocean?

PLEASE CONSIDER THE FOLLOWING:

Ships do sink. They sink until they float. No kidding. An ocean liner doesn't rest on the surface of the water—it sinks a little bit and then floats along.

To better understand how giant ships float, let's imagine a smaller ship, like a rowboat. Just like its bigger cousins, when a rowboat sits in the water, part of the rowboat sinks under the surface. Since the surface of the water used to be flat,

the boat must be pushing some water out of the way. Because the rowboat pushes water out of the *place* where the water used to be, scientists say the rowboat *displaces* water.

And we scientists know how much water the rowboat displaces. The rowboat displaces an amount of water that weighs *exactly* as much as the boat weighs. And that's true for every boat—big or small—on the water everywhere.

That's why an empty rowboat (or an empty oil tanker) floats higher in the water than a rowboat full of people (or a tanker full of oil). A rowboat with people in it weighs more than an empty rowboat. So it displaces *more* water than an empty rowboat. The boat plus the people weigh exactly as much as the water they're displacing . . . every time.

Ships float because water is pretty heavy stuff. See, an ocean liner weighs a lot and displaces a huge amount of water. But because it is *hollow*, the ship sinks only a little bit before it displaces an amount of water that's just as heavy as it is. That's why a *solid* metal ship wouldn't float. A ship made of solid steel would sink. It would displace plenty of water, but even after it displaced a ship-size amount of water, the ship would weigh *more* than the water it displaced. You know that solid metal things sink. It's because they can't ever displace an amount of water as heavy as they are.

The first guy to understand all this stuff and write it down was a man named Archimedes (ark-ih-MEE-deez). He got into a bathtub, it overflowed, and he suddenly realized that he had displaced water. He took up exactly as much room as the water he displaced. We take it for granted now, but two thousand years ago, it was a brand-new idea. The story goes

that Archimedes was so excited by this discovery that he jumped out of his bathtub, shouted "eureka" (which means "I've found it"), and went running down the street with no clothes on! Hey, it was a big deal two thousand years ago, and it still is.

Let's Sink This Chunk

You can see for yourself how a big ocean liner can sail the seas without sinking. All you need is some modeling clay and a bathtub.

- Mold some clay into the shape of a rowboat—it should look like a big clay bowl.
- Put your boat in a bathtub full of water.
- Does the boat sink or float?
- Now take your boat and squish it into a ball.
- Try floating the squished ball of clay.
- Does the ball sink or float?

When you put your clay boat in the water, it starts to sink and displaces water in a boat shape. Once the boat displaces water that weighs as much as the sunken part of the boat does, the boat stops sinking and starts floating.

The clay ball displaces water, too. But a clay ball isn't hollow like your boat. It starts to sink and keeps sinking. Eventually it displaces a ball of water the same size as the clay.

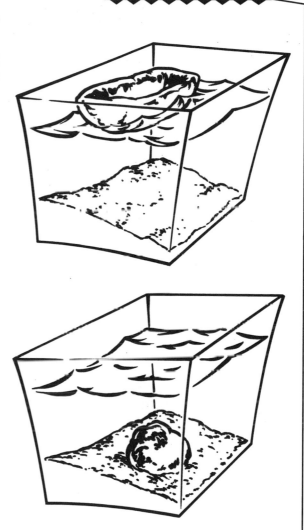

Since the clay ball weighs more than the ball of water, the clay sinks to the bottom.

Friction

Stick It to Me

You probably know a lot about friction from just walking around. If a floor is smooth, it feels slippery when you wear socks, but not if you wear sneakers. Why is that? It's the same floor, isn't it?

that you are rubbing it against. That makes *friction*, and friction makes it harder to move the sandpaper.

PLEASE CONSIDER THE FOLLOWING:

Think about something rough, like sandpaper. When you rub sandpaper against something else, the little bumps on the sandpaper hook onto the surface

Now think about a wood floor and your socks. They're both pretty smooth, so there isn't much friction. Instead of being hard to move (like the sandpaper), your socks slip and slide very easily.

But what about your sneakers? They look very smooth, smoother than socks, so why aren't they slippery, too? Well, it's because the bottoms of your sneakers are made of rubber. Rubber molecules sink into the tiny valleys in the floor and stick—more than sock molecules do.

Remember, everything that we can touch or feel is made of molecules. And buzzing around every molecule are tiny particles called *electrons*. Electrons all have the same electric charge. We say electrons are negatively charged. Other parts of molecules are positively charged. You may have heard the expression "opposites attract." It seems to be true of people, and it's definitely true of things like electrons and molecules. Not only do opposites attract but things that are alike push apart—we say they repel. There are lots of electrons in your cotton socks that repel the floor's electrons and let your foot slide across the floor. The electrons in rubber are held more tightly inside the rubber, so the sole of your sneaker doesn't repel the floor's electrons as much. That's why a sneaker sticks and a sock slides.

On the next page is an experiment that will let you reduce friction enough to make a floating vehicle!

Hover Me, I'm Going In

• •

Hovercrafts are boats that use huge fans to blow air underneath them so they can float just above anything smooth or flat, like a lake or ocean. They ride on a cushion of air. Because a hovercraft doesn't rub against whatever it's riding over, it doesn't have to drag its way through it or make waves. Hardly anything touches anything else, except for air molecules, and they don't make much friction. A boat with no friction can really cruise. You can make your very own Hovercraft of Science with a balloon, a pair of scissors, white glue, a piece of corrugated cardboard, and the plastic cap from a soda bottle.

◆ Cut the cardboard into a disk about 12 inches (30 centimeters) across.

◆ Smooth the edge of the disk with your fingers.

◆ Use the scissors to poke a hole in the exact center of your disk and the soda cap.

◆ Glue the soda cap onto the cardboard so that the holes line up.

◆ After the glue has dried, blow up a balloon and twist it shut.

◆ Holding the balloon shut, pull the mouth of the balloon over the soda cap.

◆ Put your hovercraft on a smooth surface.

◆ Untwist the balloon, and give the cardboard a light push.

The air pushes away from the cardboard and the table. That's just the push your hovercraft needs to float, or *hover*, like, well—a hovercraft!

EXPERIMENT

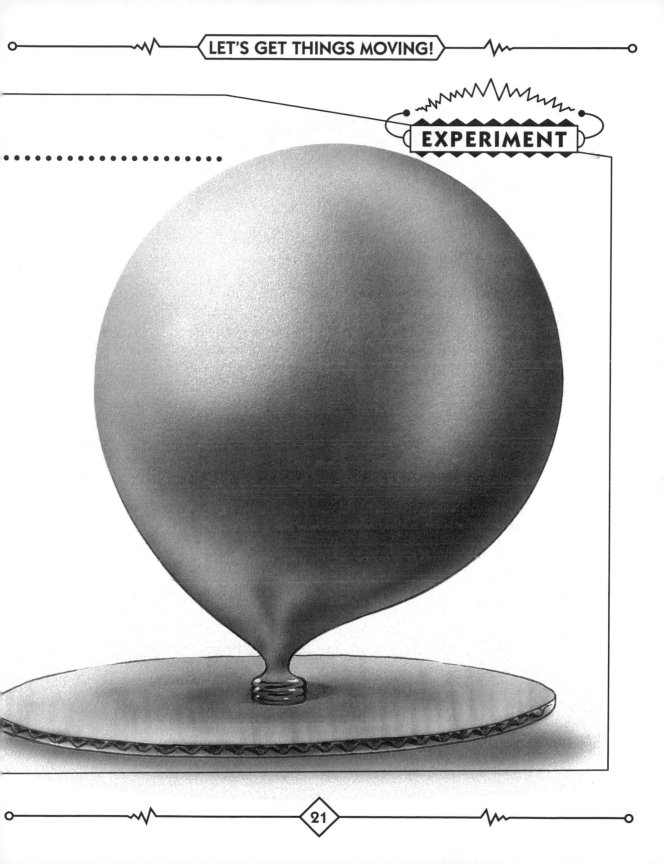

Absolute Zero

We're Talking Really, Really Cold Here

Water freezes at 0 degrees Celsius (32 degrees Fahrenheit). That's what we mean when we say "zero degrees." But scientists are often heard talking about another kind of zero. It's called *absolute zero*. What's the difference?

You know how we keep talking about molecules? That's because they're *everywhere*. They also help explain about absolute zero. You see, heat is just a form of energy. Where does it come from? Molecules. Molecules are *always* moving—vibrating back and forth, bumping into one another—just like a good bumper-car ride. You probably know that these molecules, bumping into one another all the time, make air pressure. But they also make heat.

Heat is the energy of moving molecules. See, the molecules in warm water are moving pretty fast. They move around a lot faster than molecules in cold water. You can tell by putting your hand into the water. The warmer water molecules bump into your fingers and shake up the molecules that make up your skin. By comparing the amounts of shaking going on, your nerves and your brain are able to recognize the water as warm instead of cold.

When you make something colder—say, by putting it in the freezer—you slow its molecules down. The less heat, the less energy the molecules have to do all their wiggling back and forth.

Now think about what would happen if you slowed those molecules all the way down, so they stopped moving completely. *That* would be absolute zero. It has to be pretty darn cold to stop all those molecules, though. And I mean really, really cold. Much colder than Antarctica. We're talking 273 degrees below zero Celsius (-273°C)—that's 491 below in

Fahrenheit (-491°F).

So where does it get cold enough to be called absolute zero? Well, nowhere, actually. The problem is, we can't make a freezer that gets that cold. You see, there's always some part of the freezer that touches the outside world. So even if it's close to absolute zero inside the freezer, there's always some place where heat can move through the container from the outside. And that means that no matter how hard we try, the molecules never quite stop moving. Scientists have used lasers to stop molecules from moving by bumping them from all sides. The molecules almost stop, but not quite—not for more than an instant; they stop for less time than we can measure.

So we can't ever get to absolute zero. But scientists have gotten awful doggone close—within a few thousandths of a degree.

"Cold? No, it's still 2/1000ths of a degree above absolute zero!"

Fun with Your Freezer

EXPERIMENT

Getting to absolute zero is impossible because even in the coldest freezer, the floor and the sides and the ceiling of the freezer can let in heat from the outside. To see how heat can move through objects, try this:

◆ Put an empty glass soda bottle in the freezer. Leave it there overnight, so it gets nice and cold.
◆ Once you're sure the bottle is cold, pull it out of the freezer and quickly put a penny over the bottle's mouth.
◆ Hold both your hands tightly around the bottle.
◆ Watch the penny.

The penny begins to jump up and down because the heat from your hands flows through the glass and warms up the air inside the bottle. As the air gets warmer, it expands and pushes up on the penny, making it jump. The reason the penny jumps is the same reason scientists cannot reach absolute zero in a freezer. Heat will always flow from the warmer outside to the colder inside.

Chemistry

Poison Plus Poison Equals
Something You Can't Live Without

Something we eat every day is made up of two *poisons*, yet we can't live without it.

PLEASE CONSIDER THE FOLLOWING:

Our brains and eyes and muscles must have salt to work properly. Our blood is filled with salt. Our sweat is salty, too. We're like one giant potato chip! So what's with the poison idea?

Scientists have figured out that ordinary table salt is made up of two chemicals called sodium and chlorine. *Chemical* is just a scientific word that describes a particular type of molecule. Like all molecules, chemicals have tiny particles called electrons buzzing around them. These electrons are important when it comes to salt.

See, everything in the Universe is made up of different kinds of chemicals. And the basic chemicals that make up other chemicals are called *elements*. Sodium and chlorine are both elements. By themselves, they're poisonous. It's true. If you breathed in too much chlorine, or if you swallowed a lump of sodium, you would be a goner.

But when sodium and chlorine are mixed together, the electrons in their molecules let them hook together into a completely new chemical that we can eat. In fact, we have to eat some of it for our brains and muscles to work properly. It's called *sodium chloride*, which is what scientists like you and me call table salt.

SODIUM   CHLORIDE

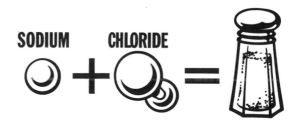

EXPERIMENT

Cleaning Pennies the Old-Fashioned Way— with Chemistry

Salt does more than just keep us alive—it's an important part of chemistry. Here's an experiment that will show you one of the cool things salt can do.

◆ Put some old pennies in a clean mayonnaise jar or soda bottle.

◆ Fill the jar about halfway with vinegar.

◆ Add a teaspoon of salt, put the lid on, and shake it.

◆ What happens to the pennies?

They become shiny and clean. When the vinegar is mixed with the salt, the sodium and chlorine in the salt don't hold on to each other as tightly. The chlorine goes after the copper in the pennies and pushes all the "dirt chemicals" out of the way. When chemicals do this kind of work, we call it a *chemical reaction.*

Rinse the pennies with water to stop the chemical reaction and keep your pennies clean. And remember what they say: "Take care of your pennies, and the dollars will take care of themselves."

Magnetism

Opposites Attract

What makes a magnet stick to a refrigerator?

them are lined up, the iron becomes a magnet.

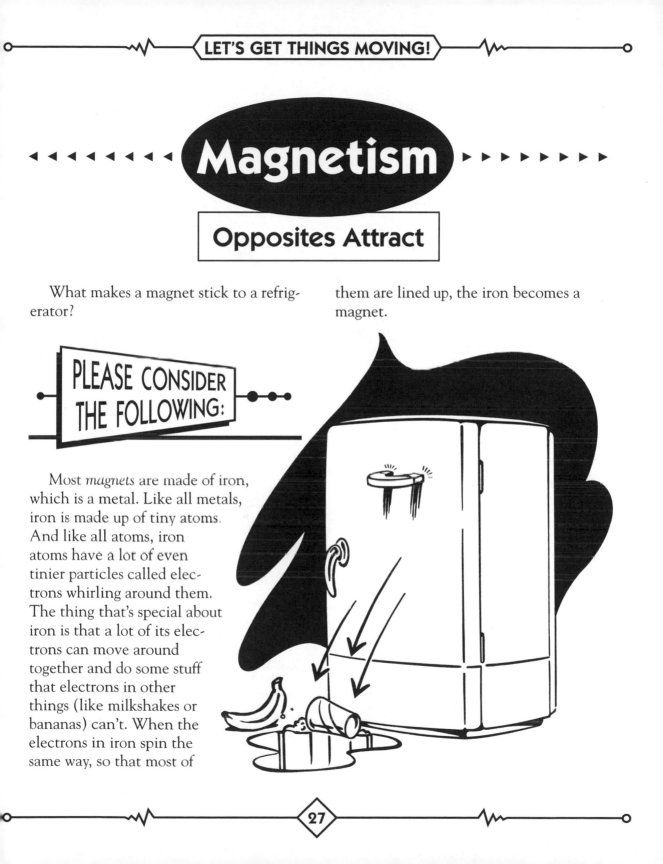

PLEASE CONSIDER THE FOLLOWING:

Most *magnets* are made of iron, which is a metal. Like all metals, iron is made up of tiny atoms. And like all atoms, iron atoms have a lot of even tinier particles called electrons whirling around them. The thing that's special about iron is that a lot of its electrons can move around together and do some stuff that electrons in other things (like milkshakes or bananas) can't. When the electrons in iron spin the same way, so that most of

When the electrons in a piece of iron line up into groups, like rows of musicians in a marching band or fields of grass blowing in the wind, the iron becomes a magnet. If you drop a magnet, the electrons can get jolted out of line, and it won't be as strong as before.

Magnets have two different sides, which are called the north and south poles. If you put the north end of two magnets close to each other, the electrons in the magnets are going around in the same direction. It's like two people trying to push a car, but they're facing each other. The car just won't go! In fact, the people push each other away. The same thing can happen to magnets. Scientists say they *repel*. Magnets can repel. (To learn more about repelling, check out the section in this book about friction.)

You've probably noticed that magnets don't always repel. They often pull together. If you put the north pole of one magnet close to the south pole of another, the magnets grab onto each other and hold tight. We say they *attract*.

Magnets are also attracted to things that aren't magnets—like refrigerators or bicycles. When you put a magnet on the refrigerator, the magnet pulls on the electrons in the steel that the refrigerator is made of. It makes them line up in the same direction as the electrons in the magnet. That lines up the electrons in the iron in the part of the refrigerator under the magnet. The magnet sticks, even though the refrigerator isn't a magnet by itself. Not bad.

magnet   **not a magnet**

Develop a Magnetic Personality—Make a Compass

Thousands of kilometers beneath our feet is the center of the Earth, which is made up of a lot of iron and nickel. As the Earth spins around, the electrons in the iron and nickel turn the Earth into a magnet. Even though the Earth is big, it's not a very strong magnet. It can't pull a metal ship to the bottom of the ocean or a hammer out of your hand. But its magnetism does go all over the Earth, and sometimes the Earth pulls tiny particles from outer space toward itself—and that's when we see the northern lights (*aurora borealis*) or the southern lights (*aurora australis*). Usually, you have to be far to the north or far to the south to see these magnetic *aurorae*. But no matter where you are, it's easy to prove that the Earth is a magnet—just make a compass.

EXPERIMENT

A compass is a magnet that always points to the North Pole of the Earth. It's easy to make a compass at home. All you need is a magnet, a sewing needle, a plastic bottle cap, and a bowl of water.

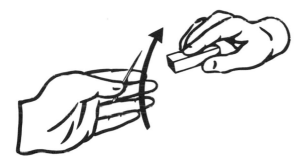

◆ Carefully float the bottle cap in a bowl of water so that it is free to move around.

◆ The needle will slowly turn the bottle cap around until it points to the North Pole! (If it doesn't work, try rubbing the needle with the magnet again.)

◆ Hold the magnet in one hand and the sewing needle in the other.

◆ Rub the magnet against the needle about 60 times. Be sure to rub the needle in the same direction every time. By rubbing the magnet against the needle, you are making the electrons in the needle line up. Pretty soon, the needle will become a magnet, too.

◆ When you're ready, lay the needle on the rim of a plastic bottle cap.

Since the Earth is a magnet, the magnetized needle wants to line up so it points to the north pole of the Earth's magnet. Now you can see where we got the idea for naming the ends of the magnet north and south. Science always points you in the right direction!

Sound

Listen . . . Listen. It's an Echo . . . Echo. What? What?

Yo-de-lay-hee-hoo! Yo-de-lay-hee-hoo! Where do echoes come from?

PLEASE CONSIDER THE FOLLOWING:

We hear echoes because sound can bounce around like a rubber ball. When you yell toward a hill-side, the vocal cords in your throat vibrate and push against the air molecules outside your mouth. The molecules bounce into one another and pass along the vibrations caused by your yell until they reach the hill. That's where the echo happens.

You need molecules to make and hear sounds. Let's say you are talking to your friends in the lunchroom. When you open your mouth to speak, you push some air from your lungs up and out of your mouth. On the way to your mouth, the air passes over your vocal cords, which are like violin strings or rubber bands stretched between your fingers. Muscles in your neck cause your vocal cords to vibrate. The vibrating vocal cords give an extra shove to the air molecules in your mouth, which pass their vibrations along to all the air molecules in the room. The vibrating air in the room jiggles your friends' eardrums so they can hear you. That's how we hear sounds.

A yell sends a wave through the air's vibrating molecules toward something like, let's say, a hillside. When the wave's vibrations hit the hill, they can't pass through it like they can through the air, so they bounce right back at you. A moment after you yell "hello!" you will hear your yell again, slightly fainter this time—"helllooo!" The time between your yell and the echo is the time it takes for the vibrating air molecules to pass their vibrations along, through the air, to the hillside, through each other, and back again. The sound goes to the hill and back at the *speed of sound*—which is about 330 meters per second (740 miles per hour).

An echo takes only a few moments to happen—because sound travels pretty fast, about as fast as a jet fighter. Whew.

EXPERIMENT

The Homemade Plastic Eardrum of Science

Here's an easy experiment to see how sounds move through the air and jiggle our eardrums.

Take a piece of plastic wrap or an old balloon and put it over a cereal bowl. Use rubber bands to keep it pulled tight over the bowl. Now you have a plastic eardrum that works exactly like your own eardrums. Of course, yours aren't made of plastic.

Whenever you hear someone talking, or singing, or a car honking its horn, your eardrum is vibrating. The same thing will happen with your Plastic Eardrum of Science. To see the vibrations, put a little salt on top of the balloon, then make some noise! Go ahead. Yell! Bang some pots togeth-er! Get loud! REALLY LOUD! See what's happening?

The sound vibrations you make move through the air and cause the plastic eardrum to bounce up and down. Notice how fast the salt shakes. It starts vibrating instantly. The sound makes it move way quicker than it would if you were to blow on the salt from a few feet away.

So anytime you hear the wind blowing or your sister singing in the shower, you'll know it's because air is moving and vibrating the eardrums in your ears. Your ear sends a message to your brain so you can figure out what you heard.
Sounds cool, huh?

Planets

Planets Come from Stars

Where do planets come from?

Planets are made of stars. No kidding. Scientists have figured out that the Earth we live on today formed billions of years ago out of an exploded star. The stars that we see in the sky at night are quite a bit like the Sun, just farther away—very far away. If you lived on a planet ten billion kilometers from Earth, one of the tiny stars you would see at night would be our very own Sun.

Stars and planets all have *gravity*. They pull on each other. Gravity is what keeps the Earth going around the Sun. And that's what keeps soccer balls from sailing off toward the Moon when you kick them—gravity.

Gravity is also what makes stars glow. Gravity makes stars crush in on themselves so hard that they explode all day and all night. And stars have some pretty huge explosions. The atomic bombs people have set off on Earth are teensy-tiny explosions compared to a star's. An average star, like our Sun, is about a million times bigger than the Earth. And the whole thing has been exploding continuously for billions of years. Those explosions make the heat that keeps the Earth warm even though the Sun is pretty far away.

As a star grows older, much older than the Sun, it starts to run out of fuel—just like when a car runs out of gas. The pressure from the explosions inside gets smaller, and the star's gravity begins crushing itself together more and more. Sometimes, there's a sudden collapse that leads to a mega-explosion, which sends bits and pieces of the star all over the galaxy. Scientists call this kind of explosion a *supernova*. The bits and pieces of the exploded star become space dust that drifts through space—and eventually forms *new* stars and *new* planets.

See, even though it's just space dust, it has gravity, too. Each piece of dust pulls on other pieces of dust. If enough dust gets pulled together by gravity, a dust ball builds up. If enough dust sticks together, the ball gets huge. Eventually the giant dust ball will crush in on itself and start exploding. That's how new stars are formed.

If there's not quite enough dust to make a star, the space dust can form a giant planet, like Jupiter or Saturn. These planets are like stars, but they're a lot smaller. They don't quite squeeze together enough to start exploding like a star.

And if just a little bit of dust sticks together, a smaller planet like Earth can be formed. Earth is pretty small compared to Jupiter or the Sun. The Earth will never become a star, because it doesn't have enough gravity.

It takes billions of years for a star to glow, explode, and then come back together to form new planets or new stars. The most recent supernova that we saw from Earth was in 1987. The one before that was in the year 1572. And we figure they happened about one and a half trillion kilometers (a trillion miles) from Earth. I guess that's what characters in the movies mean, when they talk about "a long time ago in a galaxy far, far away."

EXPERIMENT

Squeezing Planets Out of Clay

• •

Gravity makes planets into balls because it pulls in all directions at once. Try this:

Get a piece of modeling clay and start molding it. What's the shape that you might make first? Well, for me, it's always a snake, or a rope, or a string—a long thin piece. See, a snake is what you get when you squeeze the clay in two directions. You are pushing down and the table is pushing up—that's two directions of squeezing. But gravity squeezes in all directions at once. So take your clay snake and set it on end. Now squeeze it that way, too. Turn it again and squeeze it again.

After a few turns you'll see that the more ways you squeeze it, the more the clay becomes a ball. This is exactly what happens to dust in space. Gravity squeezes the dust together—in all directions at once. And that's why planets, moons, and stars are all the same shape—a shape that's a circle in every direction: a big ball. Hey, you're living on one.

Spinning Earth

It Just Keeps Going Round and Round and Round and Round . . .

How does the Earth rotate by itself?

PLEASE CONSIDER THE FOLLOWING:

About 4.5 billion years ago, tons of dust and gas drifting in space were pulled together by gravity. The dust and gas came in from all directions, but not quite evenly. Slightly different amounts of dust were pulled from different directions—randomly. So it didn't pack evenly, but started to rotate, like a whirlpool. Some of this dust clumped together into a big ball to form the Earth. As it packed in more closely, it started to spin faster.

Once you've got something spinning, it keeps spinning until something else slows it down. Scientists like you and me call this *inertia* [inn-ER-shuh]. The Earth is big and heavy, so it has a lot of inertia.

The spin is still with us today—billions of years later—because in space, there's almost nothing to slow the Earth down, to work against its inertia. So around and around we go.

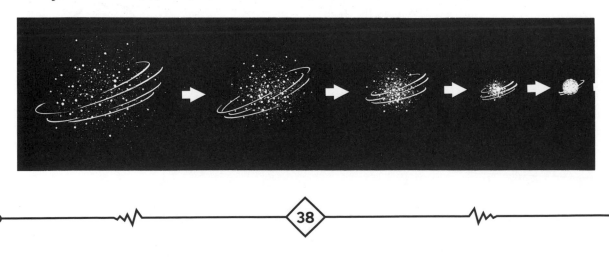

Sit and Spin

Here's a way to see how the Earth started spinning around once a day. Find a chair or stool that spins around. A lot of office chairs do this. So do some piano stools.

◆ Get a couple of phone books, like the yellow pages.

◆ Hold one book in each hand and hold your arms out straight.

◆ Sit on the chair and spin around.

◆ Move the phone books in toward you as you spin.

◆ What happens?

You'll feel yourself speed up a little. This is just what happened to the dust that formed our Earth. At first the dust was very spread out and took up a lot of space. It was probably bigger than the solar system is today. It might have had a very small spin, like the small swirls of smoke you see after you blow out a candle. It might have spun once every few hundred years. As gravity pulled the dust in closer and closer together, all its spin got squeezed into a smaller space and so it sped up. It sped up so much that the Earth now spins around once a day.

Climates

Warmer in the Middle

Why are the North and South Poles so cold?

If you look at a globe on a desk, you'll see that the world's warmest places are near the middle, not the top or bottom. The middle is where warm tropical rainforests and deserts are. It's also where wild tropical fish come from. It's hot.

EQUATOR

The imaginary line that runs around the Earth halfway between the North and the South Poles is called the *equator*. When you move north or south of the equator, it makes a big difference. Mexico, which is very close to the equator, has a warmer climate than Canada, which is farther away from the equator. *Climate* is what we call the weather in a certain part of the world—how warm, windy, and wet a place is over the course of a year.

Along the equator, the Sun stays nearly straight overhead for most of the year. Since the Earth is curved, as you move north away from the equator, the Sun will not be straight

N

S

overhead as much. By the time you get to the North Pole itself, the light (and heat!) from the Sun will just barely graze by the Earth, and it is cold—freezing cold. The same thing happens when you go to the South Pole.

That explains why closer to the equator the climates are warmer, and farther away from the equator they're cooler. Cool, huh? Or warm. Or, well, in between.

EXPERIMENT

Let the Sun Rain Down on You

You can see how climates work by shining a flashlight at a globe that's mounted on a stand or base. The Sun stays almost right overhead for most of the year along the equator. So find the equator on your globe and shine the flashlight directly on it, from about 30 centimeters (1 foot) away. Make sure the room is pretty dark, so you can see what's happening.

Still aiming the flashlight at the equator, check out the North and South poles on your globe. See how the light from the flashlight is more spread out and less intense? The Sun isn't as strong there. That's why climates are much colder at the poles than along the equator.

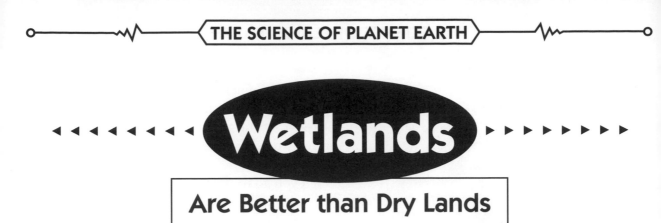

Wetlands

Are Better than Dry Lands

When there's a flood, swamps, marshes, and other *wetlands* are really good at taking in all the extra water. They're much better at it than are dry, dusty lands where it hasn't rained in months. Why do wetlands soak up water when there's a flood? Wouldn't a big area of nice dry dirt do that better?

PLEASE CONSIDER THE FOLLOWING:

A glass of water slips down your throat easily on a hot day, but water is also pretty sticky. The water you drink always goes down, because the water sticks together and gets pulled by gravity. Water sticks to water. If you take two sponges, one wet and one dry, and pour water on them both, what happens? The water will run right off the dry one but will be soaked up by the water in the already-wet one. So wetlands end up like wet sponges. They're always found where the land is lower than the rest of the land around them. So they get wet.

The plants that grow in marshes and swamps and other wetlands have roots that act like tiny dams to slow the flow of floods. Rainwater sticks to the wet mud particles in marshes, swamps, and bogs. When floods flow into wetlands, the floodwaters run into the plants and stick to the water that's already there in the mud. Floods get slowed down and soaked up by wetlands.

Dry land doesn't soak up as much water in a flood. See, floodwaters will just run right over hard, dry land—spreading the flood far and wide. Wetlands, being low and full of all their sticky water, control and contain floods better than dry lands. Wetlands are big muddy sponges.

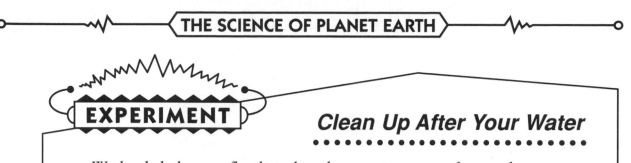

EXPERIMENT

Clean Up After Your Water

Wetlands help stop flooding, but they are important for another reason. They filter pollution out of water and help keep it clean. And that's good, because eventually water in marshes and swamps and bogs works its way into the wells and reservoirs where we get our drinking water.

To see how wetlands filter water and keep it clean, here's what you do:

◆ Add some muck like mud, pine needles, sand, or pebbles to a small pitcher of water.

◆ Stir it up, then pour the water through a coffee filter so it winds up in a glass bowl.

The water will be pretty clear! The holes in the coffee filter act like the narrow passages between the mud particles and plant roots in a wetland. In a real wetland, tiny creatures in the mud, called *bacteria*, eat the pollution that passes through. So, like a coffee filter, mud and plants and bacteria in a wetland keep water clean. When polluted water passes through a wetland, it often comes out much cleaner than when it went in. Wetlands are kind of like nature's own water filter. (But don't drink your homemade wetland, no matter how clear the water is. There are still plenty of little living things from the dirt swirling around in it that you don't want in your stomach.)

◄ ◄ ◄ ◄ ◄ ◄ ◄ # Rivers ► ► ► ► ► ► ►

Rivers Can't Run Too Straight for Too Long

Why don't rivers run straight?

PLEASE CONSIDER THE FOLLOWING:

The older a river is, the more it curves and bends. A young river—say, one that has been around for only about 10,000 years (that's old to us humans, but a 10,000-year-old river is just getting started)—runs pretty straight from mountaintop to ocean. Older rivers—ones that

are more than a million years old—tend to have more bends. We say they *meander*. The older a river is, the more it meanders.

See, rivers flow from higher ground to lower ground. They carry water downhill to the ocean. Most rivers start out as melted snow on top of a mountain or from underwater springs that bubble out of the ground. At the beginning of a river, high in the mountains, the water usually runs pretty straight. Where the land is steep, water flows straight, like a rock falling straight down. As they flow over flatter ground, rivers slow down. Bumps and hills make them change direction. As rivers flow, they slowly change where they're headed.

When the water in a river hits a big rock, it gets pushed to the side. The water is so heavy, it smashes into the banks of the river and digs out some soil and pebbles. Pretty soon, the water has carved a curve in the riverbank. That's how a river starts to meander.

Now, when soil gets knocked off a riverbank, the rushing water turns it into mud and carries it along downstream. The mud starts to fill in the bottom of the river, making the river more shallow. And shallow rivers move more slowly than deep rivers. That's why rivers tend to slow down as they get closer to the ocean. They can't move as fast downhill, because they are pushing sideways against their banks and filling themselves in with mud. The slower the river flows, the more it meanders.

That's why the older a river is, the more it curves. It has had more time to carve its banks and fill up with mud. So if you see a meandering river, it's an old one. It's been running a long time.

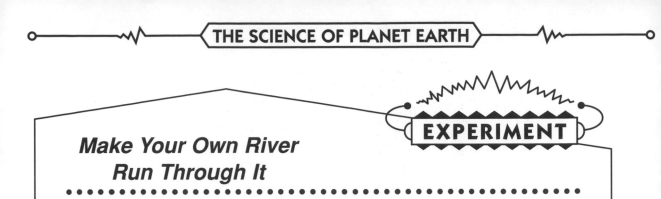

Make Your Own River Run Through It

So you don't live near a river? Well then, make one of your own. It's a snap. Start with a mountain, where a river comes from. You can build a good one outdoors with a big pile of soil, sand, pebbles, and mud. Then slowly pour a pitcher of water over the top of your mountain. Or use a sprinkler to make rain. Then stand back and watch how the water finds the quickest way down. You'll make a bunch of rivers at once. They'll carry pebbles and dirt as they move around bigger rocks. And they'll shape the land by digging into the soil. By making your own River of Science, you can make the water flow and the rocks go, just like a real river.

Garbage

It's Your Turn to Take It Out

The Earth is so big, why should we worry about how much garbage we make?

PLEASE CONSIDER THE FOLLOWING:

The average American throws away 1,600 pounds (730 kilograms) of garbage every year. We're talking 4 pounds (2 kilograms) of trash per person every single day. Multiply 4 pounds times the 250 million people in the U.S.—that's a billion pounds (one half million tons) of trash we're tossing out in the U.S. every day. Whew!

The thing is—all garbage has to go somewhere. I mean, we can't just ship it off to Saturn. It would cost too much. Most American garbage ends up at a garbage dump—or what scientists call a *landfill*. Some garbage is dumped in the ocean. Other garbage gets burned in giant ovens called incinerators. Dumping garbage in the ocean is hard on the plants and animals that live there. And burning garbage pollutes our skies. It's not so easy to get rid of garbage.

Luckily, a lot of our garbage is *biodegradable*. That means it can be naturally broken down to become soil in a few years. Things like food scraps and paper are biodegradable. But even if something is biodegradable, it still has to be exposed to air and water and microorganisms to break down. When scientists dig deep into some landfills, they find crisp lettuce and 60-year-old newspapers that they can still read. Landfills get piled so high with garbage, no air or water can get inside of them, and even biodegradable garbage can't break down.

But there are other types of garbage that definitely aren't biodegradable, no matter how much they are exposed to air and water. Most plastic, for example, is here to stay. It can't break down naturally. So as far as scientists know, all the plastic we throw away is going to stick around almost forever. (That's why it's important to recycle what you can.)

Hey, we're always going to make *some* garbage. But a billion pounds of garbage a day? That seems like a lot. We humans make a lot of garbage, and eventually we're going to run out of places to put it. Scientists realize that there's no easy answer to the garbage problem. It's something for all of us to think about.

Do Something Useful with Your Garbage: Eat It!

You can help turn your biodegradable garbage into soil. All you need is a compost bin. Here's how to make one:

Just pick out a corner of your yard where no one will mind if you make a tidy mess. Use some old boards or chicken wire to make a fence. It should be about a meter (three feet) tall, a meter and a half (five feet) wide, and a meter and a half (five feet) long. It would be best if you could build a wooden cover for your compost bin, but this isn't absolutely necessary.

Once you have your fence set up, you're ready to go. Now, instead of throwing your vegetable scraps into the garbage can, put your old potato and carrot peels, moldy onions, and dead leaves into your compost pile. Only "natural" garbage can go in the compost bin—no plastics or metal. Also, try to avoid putting meat and citrus fruit (like orange peels and grapefruit rinds) into your compost—they smell bad and attract rodents (rats and mice).

EXPERIMENT

Use a rake or a shovel to mix up your compost pile once a month or so—you want air and water to get mixed in with the scraps. In a while it should start to smell funny, but don't worry about that. After a few months, all that garbage will pack down and turn into really great soil you can spread on your flower or vegetable garden. Soil that you make yourself makes things grow like crazy!

Evolution

Let's Reach Those Trees . . .

How did giraffes get such long necks?

PLEASE CONSIDER THE FOLLOWING:

If you watch giraffes eat, you'll see

that they eat leaves and branches high up in the trees. That's what they do. They eat fresh tree salad all day.

But giraffes didn't always have long necks. Millions of years ago there were animals living on the Earth that looked like giraffes, but had short necks. They slowly changed into long-necked giraffes through a process scientists call *evolution*. We say giraffes *evolved*.

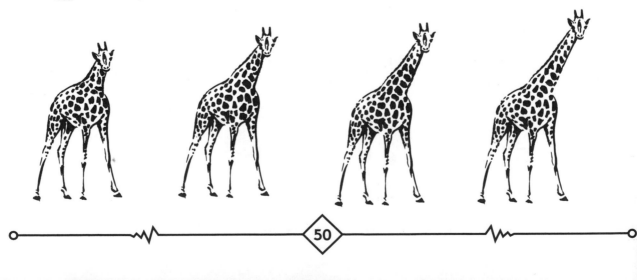

So how did those short-necked giraffes become long-necked giraffes? The short-necked giraffes had babies. And once in a while, as sometimes happens, they had a baby that was slightly different than the rest. That baby grew up to have a slightly longer neck. Giraffes born with just slightly longer necks than their neighbors were just slightly better at reaching leaves high in trees. And if you were one of those giraffes, you had a slightly better chance of always being able to eat, since there would be more for you to eat up there. If you could always eat, you'd grow stronger, and you'd have a slightly better chance of having your own long-necked kids, or "gids." (That's a joke—you know, giraffe + kid = gid. Well, maybe.)

Over hundreds of thousands of years, giraffes' necks got just a little bit longer, then a little bit longer still, and so on, and so on, and so on. That's how evolution works. So that today, giraffes have those leaves high in African trees almost all to themselves, because they have long necks, and other animals don't.

Evolution Led to YOU!

• •

Whether you are a long-necked giraffe or a short-necked human kid, all living things on Earth look like their parents (at least a little bit). That's why giraffes have long-necked kids. And no matter what you look like—whether you have curly hair or straight, brown eyes or blue—you can thank your parents and grandparents and great-grandparents and great-great-great-great-great-grandparents. Things like neck length, eye color, and hair type are passed down from parent to kid.

The blueprint for what all living things look like is carried in each and every one of our cells—in a special chemical called *DNA* (*deoxyribonucleic acid*). Almost every cell in your body contains the instructions for constructing your entire body. Half of these instructions were passed down from your mom and half were passed down from your dad. That's why you probably look a little bit like your mom and a little bit like your dad.

The best way to see how things are passed down from generation to generation is to make a family tree. The idea is to make a branching picture of your entire family. On the next page is the family tree I did for my family, the Nye family.→

You can do the same thing. Write down as much information as you can about your family in your family tree. It's amazing what you might learn about yourself.

Scientists love family trees. Except, instead of just looking at a few generations of human beings, they look back at millions of generations of all different kinds of living things. By studying the family trees of fish, spiders, pea plants, and dinosaurs, we can see how plants and animals have changed over billions of years on Earth. And that's what we scientists call *evolution*.

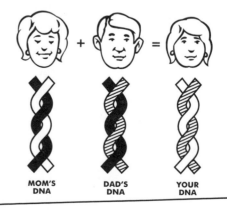

MOM'S DNA + DAD'S DNA = YOUR DNA

EXPERIMENT

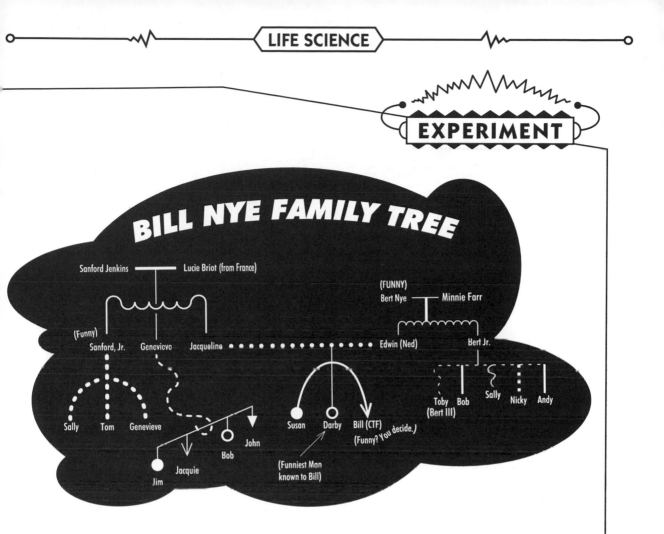

BILL NYE FAMILY TREE

A SPECIAL NOTE TO ANYONE WHO CAN'T CHECK OUT THEIR WHOLE FAMILY:

A family tree can teach you a lot about yourself, no matter who you are. The cool thing about family trees is that they don't have to be complete. Just fill in as many relatives as you can. If you are adopted or a foster child, don't sweat it. You can also try making a family tree that looks at a personality trait, like being funny. Did you get your sense of humor from your dad or your grandma? My folks say they're really not sure where I got my *particular* sense of humor.

Giant Ants

Not Now, Not Ever

Elephants are big. Dinosaurs were huge. Ants and spiders, as you know, are mostly pretty small. Why is that? Why don't ants get as big as people or cars or dinosaurs, the way they do in the movies?

PLEASE CONSIDER THE FOLLOWING:

One reason we never see dinosaur-size ants is pretty obvious. Ants are small,

and when they have babies, they have small babies, not giant "ant-asauruses." As you saw in the section on evolution, living things change (or evolve) over time to fit into particular parts of the environment. For the past 100 million years, though, ants have stayed fairly small. There hasn't been any need for ants to grow bigger and bigger to survive. Their size works perfectly for the life they lead.

But what if something happened that forced ants to make a change for the bigger? What if cockroaches moved in and began to eat all the food that the ants

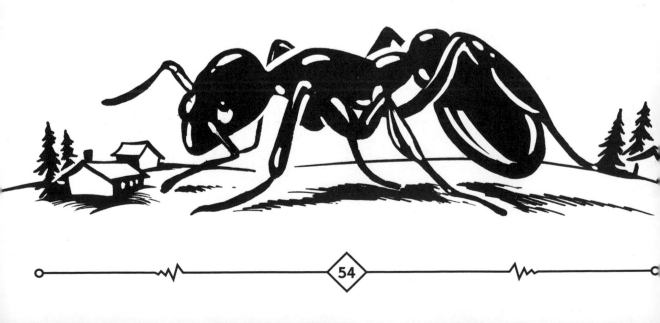

normally ate? Or what if there was a sudden increase in the number of anteaters on Earth, and the ants were being sucked up left and right? And what would happen to the ants if the oceans began to flood all the anthills on the planet? Would ants begin to evolve into giant ants, so they could eat food that cockroaches couldn't reach, or fight off the anteaters, or keep their heads above water? Probably not. And if they did, their bodies would have to change a whole lot. A giant ant with the same body as a regular tiny ant would have some serious problems.

For one thing, a dinosaur-size ant would have trouble with its blood. See, whether in an ant or a human, blood does the same thing—it brings food and oxygen to the body and takes away the heat we make when we run around. Ants and other insects don't have arteries and veins like we do to carry their blood around. Ant blood just sloshes around inside of them. This is fine for a small ant, because the blood doesn't have far to go to get to every part of its body. But a giant ant would have too much body—its blood wouldn't be able to get enough food and air around or get rid of extra heat fast enough. A giant ant would starve to death and then melt. Yecch.

There's another reason why we only see giant ants in the movies. A real-life giant ant might have to roll itself around in a wheelchair, because its legs could never provide enough support. See, most of us animals use legs for support. Humans and elephants are big, so we need thick legs to stand on. But spiders and ants are pretty small and lightweight, so their legs can be skinny and still hold them up.

Now we have to think about what would happen if we made an ant really big—say, 10 times larger than it really is. Ants and people and bricks and everything else in the Universe have three dimensions—one from top to bottom (height), one from side to side (width), and one from front to back (depth). If you make an ant 10 times larger, you have to do it in *all three* dimensions. That means you have to make it 10 times taller (top to bottom), 10 times wider (side to side) and 10 times thicker (front to back). After you multiply it by 10 three different times (once for each dimension) your new ant is going to weigh 10 x 10 x 10—1,000 times more than it did in the first place. That's one heavy ant!

And it's got to hold all of that weight up on its new, ten-times-larger-than-before legs. But there's a catch. How much weight you can hold up with a leg (or a bridge or a flagpole or anything) depends on how big it is across—what we call the size of its *cross section*. For example, a rope is as strong as its cross section. A thicker rope is stronger than a thinner rope, no matter how long the ropes are. A cross section is what you're looking at when you count the rings on a tree stump. So even though the cross section of the giant ant's leg would be bigger than it is in a normal ant's leg (10 times larger in each of two dimensions makes 10 x 10 = 100 times bigger), its size hasn't increased as much as its weight has.

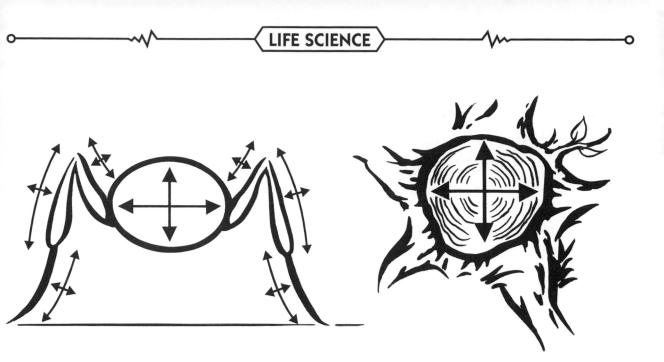

And you know what that means. In a normal ant, the legs have just the right cross section to hold up the ant's weight. If you made the ant 10 times larger, it would weigh 1,000 times more but its cross section would be only 100 times bigger. So there's just no way to have a giant-size ant—its legs would collapse. The same thing goes for movie monsters. If King Kong were a *real* ape, he wouldn't have a leg to stand on.

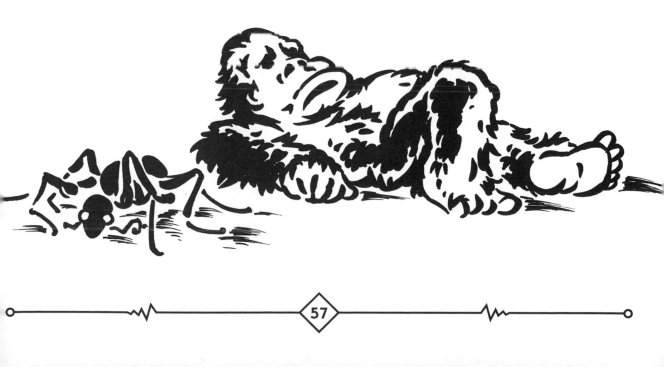

Hey, Ma—Check Out My Giant Ant!

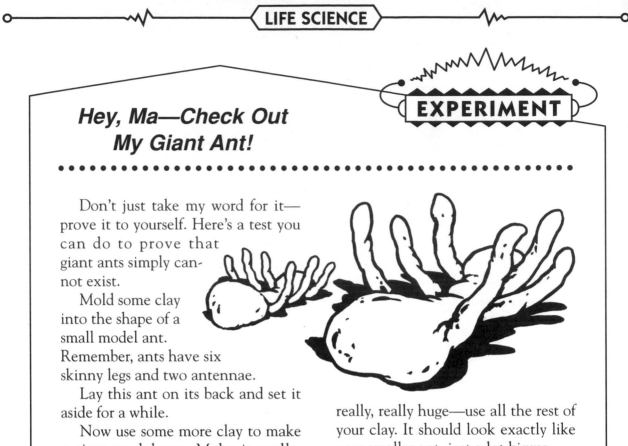

Don't just take my word for it—prove it to yourself. Here's a test you can do to prove that giant ants simply cannot exist.

Mold some clay into the shape of a small model ant. Remember, ants have six skinny legs and two antennae.

Lay this ant on its back and set it aside for a while.

Now use some more clay to make a *giant* model ant. Make it really, really, really huge—use all the rest of your clay. It should look exactly like your smaller ant, just a lot bigger.

When you are ready, turn your two ants right side up and stand them on their legs.

What happens to the small ant?

What about the big ant? Hey, don't take my word for it. Try it . . .

Reptiles

They're Cold-Blooded—Check It and See

Why do snakes hang out in the Sun?

Snakes lie out in the Sun not to get a suntan but to get the energy they need to get going in the morning. A snake in the morning Sun is doing the same thing you do when you eat your breakfast. It's preparing for the day ahead. See, snakes are what we call *cold-blooded*. That doesn't mean they're always cold. It means they're not always warm or about the same temperature, like we are. Reptiles have to warm their muscles up to get going.

If you look closely at *reptiles*—like snakes, lizards, or turtles—you'll notice that their skin is very different from your own. Reptiles have scales—small, hard

bumps all over their bodies. The scales are made out of the same stuff that's in our fingernails. Because they have scaly skin, reptiles can't sweat (you've never seen a sweaty fingernail, have you?). Sweating helps people cool off. When you run around on a hot day, sweat on your skin evaporates and carries heat away. Too much heat can be dangerous for your brain and body. So if reptiles can't sweat, how do they carry heat away from their bodies?

As it turns out, reptiles seldom need to carry heat away. Unlike humans, rep-tiles don't have to keep a constant temperature inside their body to stay healthy. See, although human bodies have to stay at around 37.0°C (98.6°F) to work prop-erly, snakes and lizards and turtles do just fine even when their body temperatures go up and down quite a bit. So, unlike us humans, they aren't set up to sweat to cool off. They don't have to be. And that's why snakes don't have to eat very much. We humans need so much heat energy, we have to fuel up and make heat by eating several times a day. But since reptiles don't need to keep a constant temperature, most snakes eat only once a month!

Scientists say that snakes are cold-blooded—not because they're mean, but because their body temperature changes with the weather. After the Sun goes down, a snake's body can get pretty chilly in the night air. And when a snake cools off, its body slo-o-ows down.

When snakes get up in the morning, after a chilly night under a rock, they need some serious heat. They need to warm up their muscles and get on with their snaky business. So they lie around in the Sun. Hey, being a reptile might not be that bad.

Humans Aren't Reptiles— No Way

As you now know, reptiles are cold-blooded. You and I, though, we're *warm-blooded*. That means our bodies stay the same temperature inside no matter what's going on outside. How would a scientist like you prove that? Just take your temperature. And it doesn't matter where you go—snuggled up in bed, out in a snowstorm, in a hot bath, or in a frigid mountain lake—your body temperature inside will stay the same, a nice and toasty 37°C (98.6°F). Humans aren't reptiles—no way.

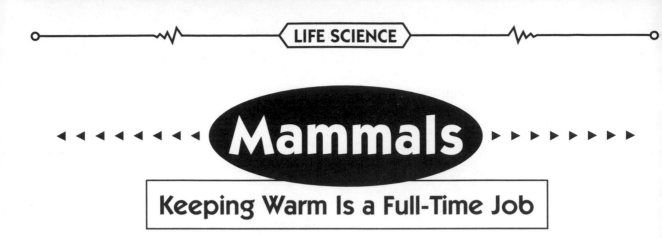

Mammals

Keeping Warm Is a Full-Time Job

Why do people take your temperature when you're sick?

PLEASE CONSIDER THE FOLLOWING:

Mammals—including horses, elephants, cats, and humans—keep the inside of their bodies the *same* temperature all the time. For humans, that temperature is usually a steady 37.0 degrees Celsius (98.6°F). Since every mammal has its own special temperature, scientists say that mammals are warm-blooded. Not because 37.0°C is so warm, but because the temperature stays the same inside our bodies.

Humans keep themselves at 37°C for a good reason—we work best at exactly that temperature. See, the human body is like those candle-powered spinning holiday toys. The heat from the candle makes them spin. They need just the right amount of heat. If you put big torches underneath one of them, it would start to spin way too fast. It might even start to wobble or fall over. On the other hand, if you blow out some or all of the candles, it slows way down or stops altogether. The same things can happen to you. If your body gets too hot (like if you got stranded in the desert), you could get sick, fall over, and even die. And if you got too cold (stranded in Antarctica this time), your body would slow down and possibly stop forever. That's why we have to stay warm all the time.

Since we're warm-blooded, we're always ready to run, jump, and especially think. We're always revved up. See, our brains couldn't work as fast at a cooler temperature. That's why scientists think that warm-blooded animals are smarter than cold-blooded animals. Of course, it's hard to tell. I mean, you can't make a snake sit down and take a geography quiz. But one thing's for sure—warm-blooded animals have the *biggest* brains of all the animals on Earth. So keeping our insides warm seems to be the reason we mammals are so smart.

The human body is well designed to keep the temperature at exactly 37°C. All mammals have hair, to help us stay warm when it's cold outside. We all sweat, to cool off when we get too hot. Another thing we have to do to stay warm is eat. Our bodies get energy out of the food we eat, and use that energy to keep warm. Since we have to keep that warmth going all the time, we have to eat three times a day.

When we get sick, our bodies sometimes turn up the heat a little bit. This alerts the doctor that something is not right, but the extra heat also helps fight off the germs that are making us sick. A temperature of 38°C (100.4°F) means you're a little sick; 40°C (104.0°F) means you're really sick, and your body is dangerously overheating. But keeping things at 37°C is just right. It lets us move fast and think fast; we're *always* warmed up.

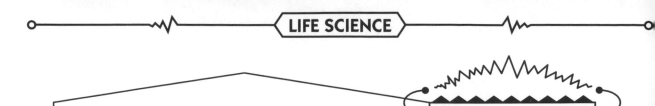

A Hairy Demonstration

We mammals all have hair or fur to keep us warm when it's cold outside.

Here's a way to see how hair does its job. Fill two glasses with hot tap water. Wrap one glass in a dish towel or washcloth. Then put each glass in a bowl of ice water. Wait 10 minutes, then test the water in the glasses with your fingers. The water in one glass has gotten kind of chilly, hasn't it? And the other is still pretty warm. That's because the glass wrapped in a towel is just like a mammal. The glass has hair—well, a towel—to keep it warm. Science is cool . . . I mean warm . . . I mean . . . well, you know what I mean!

Ocean Life

Blue on the Top and Black on the Bottom

When you look straight down into the ocean, you can't see very deep. And, if you were down there, you could hardly see anything at all.

PLEASE CONSIDER THE FOLLOWING:

It's hard to live where there's no light. And the bottom of the ocean is a pretty dark place. See, water soaks up light. Near the surface of the ocean, near where the Sun shines, there is a lot of sunlight. As you go deeper there's a lot less. In fact, on the deep ocean floor there is no sunlight at all.

The top part of the ocean, where there's a lot of light, is called the *photic zone*. That means "zone of light." Most of the life in the ocean lives there. And every ocean plant lives there. Plants in the ocean don't need to grow in soil, like plants on land. They just float around held up by the sea, soaking up sunlight.

The most common plants in the ocean are so small you can't see them with just your eye. You need a microscope. These types of plants are called *phytoplankton*. That means "plant drifter."

There are also many tiny animals that feed on the tiny plants. They're called *zooplankton*, like animals that would drift at some wild water zoo. There's a lot of plankton. If you weighed all the living things in the ocean on a giant scale, including the whales and sharks and giant tunas, the tiny plants and animals would make up 90% of the total. That's a lot of plankton!

Plankton is what scientists like to call the group of tiny drifting little things. Each little plant or animal is called a plankter. Anyway, plankton needs the Sun to survive. In fact, almost all life on Earth depends on the Sun. The Sun provides the energy for green plants (like trees, grass, phytoplankton, and seaweed) to make their own food. Plants don't really eat; they use sunlight to make their own food. So, where there's no sunlight, plants can't live. That's why plankton lives near the surface, in the photic zone. By the way, a huge amount of the oxygen in the air we breathe is made by phytoplankton way out in the ocean. The ocean is so huge and there are so many phytoplankters that we can take deep breaths. Almost every living thing depends on plants in some way. For instance, humans eat vegetables, cows eat grass, and deer eat leaves. Even animals that prey on other animals—like tigers and lions—still depend on plants.

The animals *they* eat are usually plant eaters.

Now, the same thing is true in the ocean. Small fish eat the plankton. And then bigger fish eat the smaller fish. It's a *food chain*. And plankton is the first link in the chain. All the animals in the ocean depend on the Sun to keep the plankton alive.

So the deeper you go in the ocean, the less light there is, and the less plankton you'll find. The animals that live in the pitch-black bottom of the sea eat dead fish and seaweed that fall down from the photic zone. Or they eat one another. Either way, there's not nearly as much to eat down there as there is higher up in the ocean, where there's lots of sunlight. And that's why we usually don't see nearly as many living things near the bottom of the ocean as we do at the top.

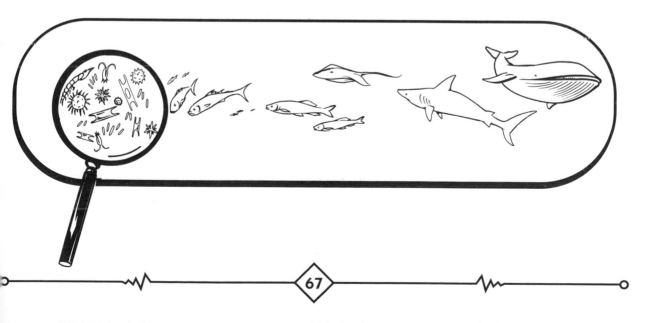

EXPERIMENT

Hey, Who Turned Out the Lights?

Almost every living thing needs the Sun to survive. That's because the Sun keeps plants alive—and we all depend upon plants for food. Here's an easy experiment you can do to see how much plants need the Sun.

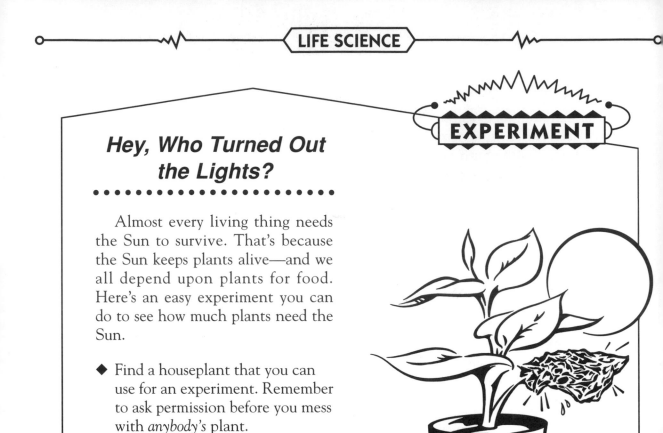

- ◆ Find a houseplant that you can use for an experiment. Remember to ask permission before you mess with *anybody's* plant.
- ◆ Cover one leaf *completely* in aluminum foil. Cover the stem attached to the leaf, too.
- ◆ Leave the plant where it will get lots of Sun.
- ◆ In a week, remove the aluminum foil and look at the leaf.

Your leaf has probably become withered and yellow. The aluminum foil prevents light from getting to the leaf, so the plant shuts down. The yellow color is a sign of a dying leaf. And that's why there are fewer things living on the bottom of the ocean. The water above the ocean floor acts like a giant piece of aluminum foil and keeps sunlight from getting to the bottom. No light means no plants, and no plants means no food for fish to eat. Almost all life, even life on the ocean floor, depends on the Sun.

Fish

Where, oh, Where Is the Air Down There?

Where do fish get the air they breathe?

PLEASE CONSIDER THE FOLLOWING:

Fish breathe air, just like we do. Air is invisible, but it contains many different chemicals that are important to living things. *Oxygen* is a chemical in air that helps us get energy out of the food we eat. All animals—including fish—need oxygen to survive. And fish get most of their oxygen from the same place we do—plants.

See, plants don't need oxygen to live. In fact, not only don't plants need oxygen, all plants actually *make* oxygen every day. It's kind of like, when plants exhale—or breathe out—they exhale oxygen. So we humans breathe in what plants breathe out.

Fish do the same thing. They get the oxygen they need from plants in the ocean. Phytoplankton and seaweed are plants that make oxygen all the time. But since these plants live underwater, the oxygen they give off gets mixed right into the water, where fish can breathe it.

Fish get most of their oxygen from plants in the water, but they can also get it from the air that's right above the water they live in. On Earth, water gets churned around all the time. Air mixes with water when it rains or when wind blows over the ocean or a lake. It happens all the time in mountain streams, where rocks and rapids make the air bubble into the water. You see it when a stream has "white water" or when an aquarium pump makes bubbles. They're not just making noise—the bubbles let fish get some air.

Even though all animals breathe oxygen, we humans can't breathe the oxygen fish breathe, and most fish can't breathe the oxygen we breathe. You and I have two lungs, which do a great job getting oxygen out of the air and into the blood. The blood constantly carries oxygen to every part of the body. But most fish don't have lungs, so they can't get oxygen out of the air and into their blood like you do. Instead, fish have structures called *gills*, which are specially designed to take oxygen out of the water. Gills are the red fan-shaped things you see if you look closely at a fish underwater. Behind a flap called the *operculum* (oh-PERK-you-lumm) on the sides of their bodies, you can see the gills. They're red because they're filled with blood. In the gills, oxygen from the water passes right into the fish's blood. So if *you* ever want to breathe underwater, grow some gills . . . or bring a scuba tank and use your lungs . . . hmm.

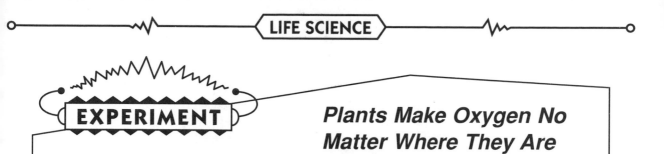

EXPERIMENT

Plants Make Oxygen No Matter Where They Are

Humans and fish both need oxygen to breathe. Luckily, plants make oxygen all the time. They take in a chemical from the air that we humans breathe out—called *carbon dioxide*—and give off oxygen. It's an extremely convenient relationship. It doesn't matter if a plant lives on a mountainside or below the ocean surface, plants make oxygen whenever there's enough sunlight. Here's an experiment you can do to see a plant making oxygen. You'll need a leaf of really fresh lettuce, a large glass bowl of water, a piece of cardboard, and a glass jar (like a mayonnaise jar) with the lid removed, filled nearly to the top with water.

◆ Put your lettuce into the water-filled jar.
◆ Then put the cardboard on top of the jar and hold it in place like a lid.
◆ Still holding the cardboard, turn the jar upside down and put it into the bowl of water. (It's OK if the cardboard gets wet.)

◆ Once the jar is in the glass bowl, pull out the cardboard. The water and the lettuce should stay high up in the jar.

Put the jar and the bowl where it will get plenty of light. In a couple of hours, you'll start seeing bubbles. Those bubbles are filled with oxygen. Green plants make the oxygen that keeps us humans going.

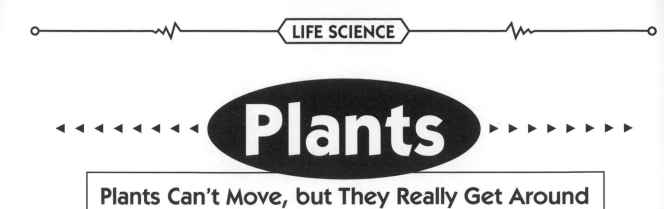

Plants

Plants Can't Move, but They Really Get Around

Why do apple trees make apples?

PLEASE CONSIDER THE FOLLOWING:

Apple trees need animals to plant their seeds for them. See, apple trees are plants. Like all plants, they can't really go anywhere on their own. They don't have legs or bicycles—they're stuck. To get around this, many plants make a delicious fruit for animals to eat. And when we animals gobble down apples, we give their seeds a lift to another piece of land. Then the seeds have a chance to grow. That way, the apple tree can make more apple trees without actually having to go plant them.

Plants and animals spend a lot of time and energy making new family members. If they didn't, there would never be any new plants or animals. Now, if you're a tree (and I'm not saying you are), you wouldn't want all your kids—your young trees—growing right underneath you. You'd have too many kids living too close. You'd all be trying to use the same patch of land and the same patch of sunlight. So you have to spread yourself out a little.

Since you're a plant, you can't move on your own. You can't walk around the neighborhood planting your seeds and watching them grow. But if you're an apple tree (or a pear tree or an avocado tree), you can do the next best thing. You can wrap your seed in a fruit and let animals do the walking for you. Hey, apple trees seem pretty smart.

The Hitchhiker's Guide to the Plant World

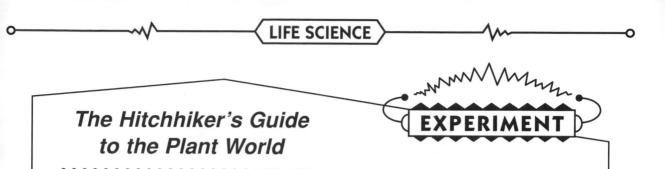

EXPERIMENT

• •

Even though plants can't move, they've figured out lots of different ways to get around. We've already seen how fruit trees use animals to carry their fruits (and seeds) around. But plants don't need to make fruit to move. Dandelions turn their flowers into fuzzy parachutes. When the wind blows, little dandelion seeds float through the air to be planted somewhere new. Maple trees have little "helicopters" that float their seeds away from the tree. And plants that live near rivers often have seeds that can float to a new home.

Another way plants can move around is by sticking to animals that brush by them. Anytime you walk in the woods, you probably carry seeds around without even realizing it. Here's a way to see how plants hitch a ride on humans and other animals.

◆ Find an old pair of white socks and a magnifying glass. The larger socks you can find, the better.

◆ Put the socks on *over* your shoes and socks.

◆ Then go for a walk through the woods or a field.

After about half an hour, take off the socks and look at them carefully with the magnifying glass. You will find little burrs and barbs stuck in your socks. These are seeds from the weeds you've been walking through. The same thing that happened to your socks also happens to animal fur. Instead of using a fruit, these plants have little "hooks" to hitch a ride from place to place. That way, they're sure to keep on growing. They'll find a place; they're weeds.

Trees

Always Turning Over a New Leaf

Why do some trees have wide leaves while others have skinny needles?

PLEASE CONSIDER THE FOLLOWING:

Trees have different kinds of leaves depending on where they live in the world. Trees with wide, thin leaves usually live where the Sun shines brightly for at least part of the year. That way, their broad leaves can soak up a lot of sunlight. Remember, all plants (including trees) use sunlight to make food and survive. So trees with big leaves can make a lot of food whenever the Sun shines. They store up the food they make. Then, every fall, these trees let their leaves fall off. (Get it? The leaves fall in the fall.) These types of trees are called *deciduous* trees. *Deciduous* is a word that means "fall off." Deciduous trees lose their leaves before it gets too cold or too dry. The water in their leaves would get sucked out by dry air. And wide, thin leaves can't survive weather that's too cold. They would freeze solid and break off. That's why deciduous trees shut down food production every fall. The tree absorbs all the water and food from its leaves and then lets them go—lets them fall. (Ha.) In springtime, the tree will grow a whole batch of new leaves. No problem.

But *evergreen* trees—like pines and firs—don't have wide leaves. They have thin, sharp needles. These needles are tough—much stronger than your typical deciduous tree leaf. But needles do the same thing that big leaves do—they help a tree make food. Evergreen trees need tough needles, because they live in places where it's really cold or really windy or in places that don't get a lot of sunlight year-round. And the needles have a waxy covering, so they don't dry out very fast.

Evergreen trees keep their needles for three or four years. As evergreens grow new needles, they shed the old ones. Not all at once, in the fall, like deciduous

trees. Instead, evergreens lose their needles all the time as they make new ones. They keep a constant supply of needles so they're always prepared to make food when the Sun *does* shine.

EXPERIMENT

Trees Get Thirsty Too, You Know

Trees need a lot of water to live and grow. And since they don't have blood, like you and me, they use water to move things around inside themselves. Some trees need to take in a *ton* of water a day—that's 1,000 kilograms (more than 2,000 pounds)—as much as a small car weighs! And when you stop and think about it, you can imagine that water has to go a long way to get from the ground to the top of a tall tree. All plants have a system of tubes and tunnels inside of them that helps carry water and food around.

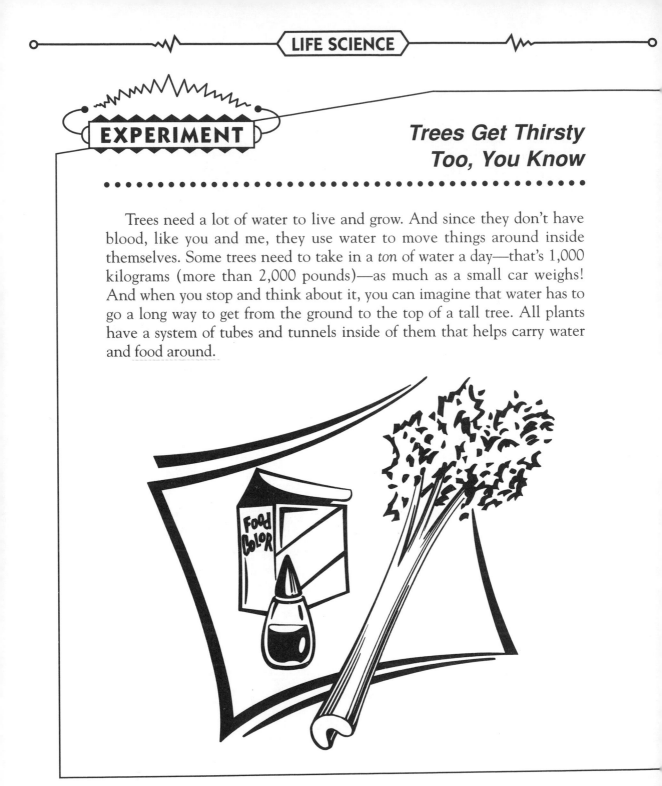

Plants don't suck water up out of the ground like a straw. Instead, they use water's tendency to stick to things to move it up. It's the same thing that makes your clothes get wet a little higher than the water level when you step in a deep stream or stand where ocean waves hit you. It's called *capillary action*. You can see how it works. All you need are a few stalks of celery (with the leaves left on), food coloring, and three tall drinking glasses.

◆ Fill the drinking glasses with water and color each one with five drops of food coloring. Put a stalk of celery in each glass.

◆ Set the glasses somewhere in the house, *out of* the sunlight.

◆ In a day, you'll see color in the leaves.

All plants (like celery and trees) use sunlight to make their own food and stay alive. So during the day, when the Sun is shining, they concentrate on making a lot of food in their leaves. After the Sun goes down, they drink up lots of water from their roots. That's why you have to keep your celery out of the sunlight for this experiment to work. Now try putting your glasses of celery and food coloring in the bright sunshine. What happens?

Leaves

So Many Beautiful Colors

Why do leaves turn red, yellow, orange, and brown in the fall?

PLEASE CONSIDER THE FOLLOWING:

When trees get ready for winter, their leaves take a break from being green. See, tree leaves all contain an amazing chemical called *chlorophyll*. *Chloro* is a Greek word that means "green." There's lots of chlorophyll in leaves, and it's chlorophyll that gives them their green color. Chlorophyll lets plants (like trees) make their own food from sunlight, water, and carbon dioxide. We humans have to eat food to get energy, but plants create special types of sugar inside

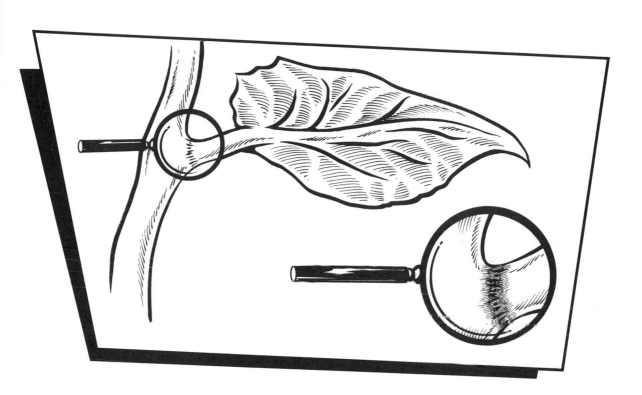

their cells whenever there's enough sunlight. The sugar provides energy for plants to grow.

In the fall and winter, there's less sunlight. The days are shorter and the nights are longer. A lot of trees take a break from making both chlorophyll and food. They store up enough sugar in the spring and summer so they don't have to worry about starving to death in the winter.

Leaves are sensitive to light, so they can tell when the days start to get shorter. That's why, every fall, deciduous trees grow a thin layer in the stem of each of their leaves. This layer is called the *abscission* (ab-SIH-zhun) *layer. Abscission*

is a word that means "cut off from." The abscission layer is so brittle that just a slight breeze will break the leaf off the tree. It also stops the leaf from making any more green chlorophyll.

When that happens, all the other chemicals in the leaf start to show through. There's stuff like *xanthophyll* (ZAN-thoh-fill) and *carotene* (CAHR-uh-teen). *Xantho* means "yellow," and carotene is what makes carrots orange. These chemicals are in the leaf all year long, but you can't see them when there is a lot of chlorophyll. In the fall, leaves stop making chlorophyll, and the leaves' other cool colors show through.

See Some Chlorophyll

Chlorophyll is a natural chemical that gives plants their green color and lets them make their own food. Here's an experiment you can do to see chlorophyll and the other colored chemicals in plants. You'll need plant leaves, measuring spoons, a tall drinking glass, clear nail polish remover (also called *acetone*), a coffee filter or blotting paper, scissors, and a pencil or chopstick.

◆ Collect a few green leaves from a tree.

◆ Tear the leaves into very small pieces, like confetti.

◆ Use a spoon or your thumb to break up the leaf pieces until you have about a teaspoon of plant mush.

◆ Then put your mush in the bottom of a tall drinking glass.

◆ Slowly add about five teaspoons of nail polish remover to the glass.

◆ Wait about 20 minutes to let the plant mush sink to the bottom of the nail polish remover.

EXPERIMENT

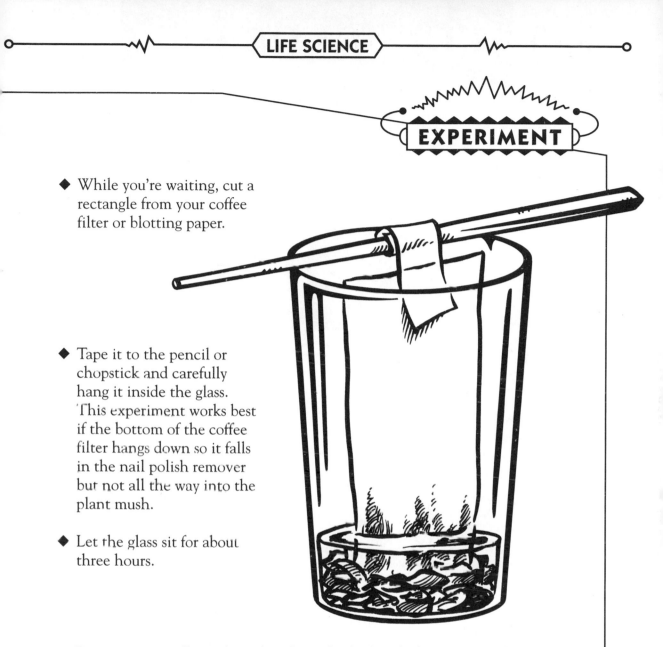

◆ While you're waiting, cut a rectangle from your coffee filter or blotting paper.

◆ Tape it to the pencil or chopstick and carefully hang it inside the glass. This experiment works best if the bottom of the coffee filter hangs down so it falls in the nail polish remover but not all the way into the plant mush.

◆ Let the glass sit for about three hours.

Pretty soon you'll see the colors from the leaf work their way up the strips of paper and spread out. These are the same colors you see in leaves. The green is chlorophyll and the other colors are the same chemicals (like xanthophyll and carotene) you see when leaves change colors in the fall.

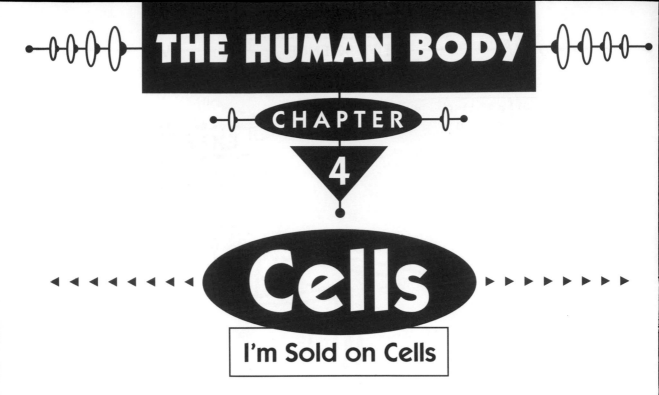

Cells

I'm Sold on Cells

If the human body is made up of cells that are too small to see with just your eye, what are cells really like?

PLEASE CONSIDER THE FOLLOWING:

If you think of the human body as a house, then a cell is like a room in the house. Since the human body is made up of more than a hundred million million cells (100,000,000,000,000), we're talking about a pretty huge house with a whole lot of rooms, but the idea is the same. Depending on what they're used for, rooms can look very different from one another. But most rooms in a house have a lot of the same things—walls, floors, ceilings, doors, and windows. A closet and a kitchen are very different, but they're still both called rooms. The same thing is true for cells—a brain cell and a skin cell don't look very much alike, but they still have a lot in common.

Like a room, a cell has walls, which separate it from other cells. Plant cells have cell walls that are thick and strong. This gives plants, like trees, their strength and structure. Animal cells, like the cells that make up you and me, have cell *membranes* instead of cell walls. *Membrane* comes from a Latin word that means "thin skin." And a cell membrane is just a thin, flimsy wall that can bend and bulge easily. Most animals need to

move around, and a cell membrane provides more flexibility than a stiff cell wall.

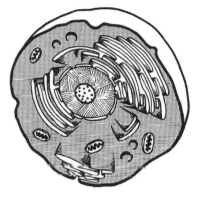

ANIMAL CELL

PLANT CELL

In an office, you'll probably find a bunch of file cabinets filled with papers. They might be filled with information. In a cell, you'll find a *nucleus*. It looks like a small dark blob inside the cell. The nucleus is the cell's information storage center. It's the place where the cell stores its own blueprints. Growing, multiplying, moving around, and even dying—the instructions for everything a cell does are stored in the nucleus.

A kitchen usually has a refrigerator where you can find food to eat. Food gives us energy to live on. Well, the same thing is true in a cell, except instead of a refrigerator, cells have tiny structures

called *mitochondria* (mite-oh-CON-dree-uh), where the cell makes the energy it needs. If a cell's mitochondria stop working, it will get tired, slow down, and eventually die.

A grocery store is a giant room with lots of shelves. At the front of the room, there's a checkout stand where your groceries are bagged up. Cells also have a checkout stand inside them—a structure called the *Golgi* (GOAL-jee) *body*. The Golgi body collects chemicals inside the cell and packages them up together in small membranes. The Golgi body then pushes these minimembranes to other parts of the cell or even out of the cell so

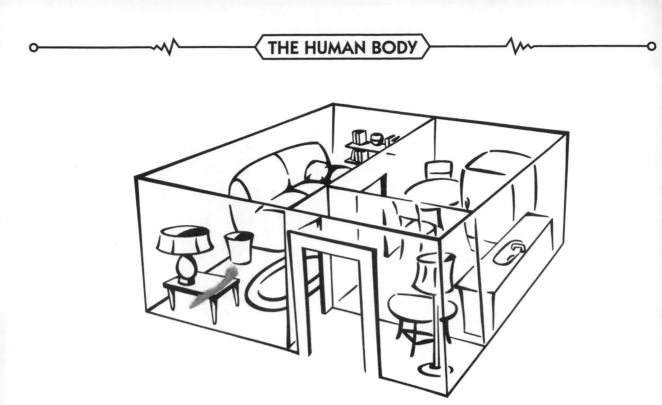

they can travel to other parts of the body. Two types of chemicals that help our bodies grow and work, called *hormones* and *enzymes*, are often bagged up by the Golgi body.

In most rooms, you can find a wastebasket where you throw away scraps and papers. A cell has a wastebasket, too. *Lysosomes* (LIE-suh-soams) are sacks of powerful chemicals that help break down the food we eat and any poisons that find their way into our cells. Cells also have things that rooms *don't* have. Like the *ribosomes* (RYE-bo-soams), where cells make protein. Almost everything in your body is made of proteins. Plus, cells are filled with watery stuff called *cytoplasm* (SITE-oh-plaz-um). The cytoplasm contains a lot of water and helps the cell move food and energy around. A room filled to the top with watery cytoplasm wouldn't be a room, it'd be a swimming pool!

And just as there are different kinds of rooms—kitchens, bedrooms, and bathrooms—there are different kinds of cells: brain cells, bone cells, blood cells, and skin cells, just to name a few. Each one has its own special look. Nerve cells are long and thin to carry messages around your body. Blood cells are round and flexible, to squeeze through the spaces between other cells. Muscle cells let you move because they are stacked on top of one another. And, the stacks can shorten themselves. They're all different, but they have plenty in common—they're all cells!

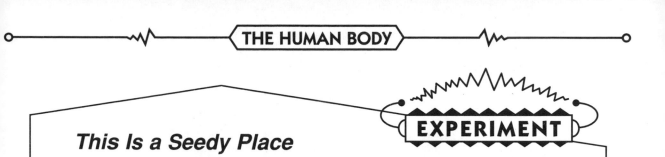

EXPERIMENT

This Is a Seedy Place

Look inside most cells, and you'll find a lot of water. There's so much water in cells that we humans end up being more than two-thirds water. But the cells in plant seeds are pretty much dried out. The stuff inside their cells—the cytoplasm—loses a lot of water while the seeds wait for the weather and the soil to become just right for growing.

For the plant cells in a seed to start growing and make a plant, their cells need water and sunlight. You can see how this works with a handful of dried beans. You'll need hard, dried beans—any kind will do—black beans, red beans, black-eyed peas, even dried garbanzo beans (chickpeas); a cereal bowl; and cotton balls.

◆ Fill the bowl with cotton balls and mix in about 15 of your dried beans.

◆ Add just enough water to get the seeds wet—they don't have to go swimming!

◆ Then put the bowl in a sunny window.

Over the next few days, the seeds will soak up the water and get bigger. The seeds will come alive and grow baby plants! You can transfer the seeds to your garden or a flowerpot. What started out as dried-out plant cells will become a living, growing, water-filled bean plant!

Germs

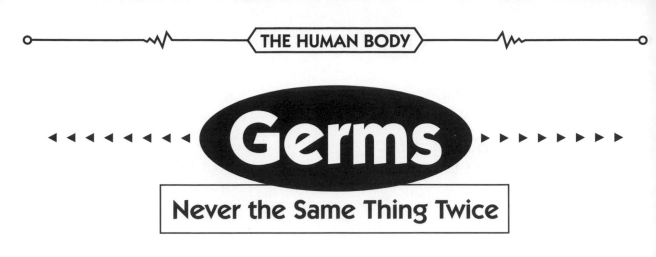

Never the Same Thing Twice

If the human body can fight off almost every germ that gets inside it, then why do so many people get a cold *every* winter?

PLEASE CONSIDER
THE FOLLOWING:

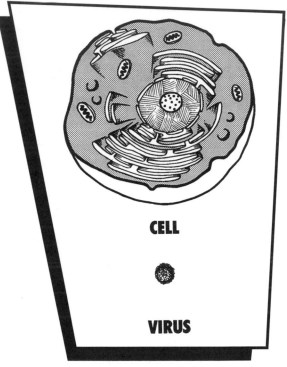

CELL

VIRUS

We keep getting colds because the germs we're trying to fight off keep changing. The germs that cause colds are called *viruses*. Viruses are tiny, tiny chemical constructions that don't drink, don't eat, and don't dance. They are the tiniest living things we have found on the planet. If this is a cell, this is how big a virus is. Actually, it's hard to say if viruses are alive. They seem to be a bridge between living things and chemicals. They don't reproduce on their own. Viruses get inside another living thing's cells and make copies—of themselves. After a

while, all these newly copied viruses can burst out and destroy the cell; then they can move into some new cells. Oh, it's not good.

Now, during all this, the viruses cause the cells they've entered to put out a special chemical—a protein—on their surface. And that's a good thing, because this protein sends a signal out to other cells that a virus is inside. It's kind of like Halloween, when you turn on your porch light as a signal to the trick-or-treaters to come get some candy.

When special cells in your blood—your white blood cells—notice this protein "porch light," they begin to multiply like crazy. Actually, you've got lots of different kinds of white blood cells, and just some of them—the ones that can connect to that signal protein and kill your virus-infected cells—start to multiply. When the white cells go to work, you sometimes feel sick or have a fever. After a few days, if you're lucky, you've made enough white blood cells to kill off the cells with viruses inside them. When your white cells kill the cells and destroy the virus, you start to feel better. And the next time that virus comes along, your white blood cells can act faster and keep you from getting as sick. Since you've already fought it off once, you and your blood cells are ready.

If a cold virus always stayed the same, you would never have that particular kind of cold again. But sometimes when a virus is in your cells making copies of itself, it makes a copy that is slightly different from the rest. This new virus causes the cell to make a slightly different signal protein than before. Most of the time, the white blood cells can handle it. If your white blood cells can't recognize the signal protein, you might start getting sick all over again. Then you have to wait until the right kind of white blood cells notice the new protein, multiply enough to kill the cells, and wipe out the new, slightly different copies of the original virus. Whew!

There are a few viruses that attack the very things we use to fight off viruses—our white blood cells. That's what HIV (Human Immunodeficiency Virus) does. If HIV invades the body, people can get sick from other diseases that normally would be no problem. That's the Acquired Immune Deficiency Syndrome (AIDS). HIV is one dangerous virus!

Sometimes a new virus is made inside one person, who passes it along to another. Viruses can travel around the world, just by people sneezing and shaking hands. As a virus moves from person to person, it makes copies of itself and changes. By the time a virus has worked its way back to you, it can be very different. Then you're *really* not ready for it, and you get sick—again.

Other germs, like viruses, are always changing and our bodies are always learning how to fight them. It's just a part of life. So if you want to help your body fight germs and stay healthy, don't eat or touch your mouth or eyes without washing your hands first. No kidding. Soap and water help carry germs away from your hands before they can find a way inside your body!

EXPERIMENT

Grow Your Own Germ

It's pretty tough to see germs because they are so small. We can usually see germs only through a microscope. But try this and you'll see how germs can multiply into a whole colony of bacteria or mold. You'll need three tall drinking glasses, vinegar, salt, and three cubes of beef bouillon.

◆ Fill all three glasses with tap water.

◆ Drop a bouillon cube into each glass. Wait three minutes and then carefully stir a teaspoon of vinegar into one glass.

◆ With a clean spoon, stir a teaspoon of salt into the second glass.

◆ Leave the third glass alone. Put all three glasses in a sunny window for five days. What happens?

You will probably find large colonies of germs growing in the glass that has no vinegar or salt in it. The water in the glass will probably be cloudy—filled with millions of bacteria too small to see without a microscope. There might be some green and white mold growing on top of the water. Mold comes from microscopic creatures that live in the air and wait for a place to settle down and grow. The beef bouillon provides a great home for bacteria and mold, so they both grow like crazy.

But what about the vinegar glass and the salt glass? Those should be clear. Vinegar and salt prevent germs from growing. And that's why things like pickles, kimchi, sauerkraut, and beef jerky are safe to eat even though they aren't usually cooked with heat. Instead, pickles, kimchi, and sauerkraut are made with cucumbers or cabbage soaked in vinegar. The vinegar changes the cucumbers and cabbage and gets rid of any germs that might be around. And beef jerky is made from salted beef. The salt kills bacteria before it can grow.

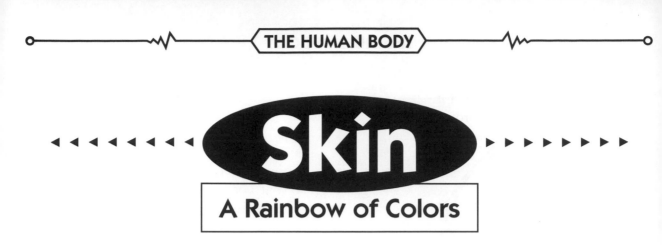

◀ ◀ ◀ ◀ ◀ ◀ ◀ **Skin** ▶ ▶ ▶ ▶ ▶ ▶ ▶

A Rainbow of Colors

Why do some people have darker skin than others?

PLEASE CONSIDER THE FOLLOWING:

All of us have brown skin. Some of us have dark brown skin, some of us have light brown skin, and some of us just have brown freckles in our skin. We're one brown bunch of humans. But no matter how brown you are, your "brown-ness" comes from a chemical in your skin called *melanin* (MEL-uh-nin). Melanin is a dark-colored chemical, so the more melanin you have, the darker your skin will be.

Melanin gives us color, but it also pro-tects us from getting too much Sun. It acts as a natural sunblock. We have to let *some* sunlight into our skin. Sunlight helps our skin make an important vita-min—vitamin D. We need vitamin D to keep teeth and bones healthy.

So when you get out in the sunshine, you make vitamin D and get darker or tan. The melanin in your skin acts like a sponge. It can hold a lot of the Sun's energy. But if a sponge gets too full, it can't hold any more. Too much sunlight can overload our melanin and give us a sunburn. That's why people with really light skin sunburn easily. They don't have as much melanin in their skin to protect them. (Although melanin does a good job of protecting your skin from too much Sun, it has its limits. That's why it's important to use a sunblock whenever

you're out in the sunshine for a while. If you help your melanin protect your skin from too much Sun, your body will stay healthier and happier.)

Since each person has a slightly different amount of melanin, we each have a slightly different skin color. There's brown, pink, olive, and yellow, just to name a few. Now, how much melanin you have in *your* skin depends on how much your parents have in theirs. The instructions for how much melanin your body makes is stored in the DNA inside all your cells. Half your DNA came from your mom and half came from your dad. So, people's skin color depends on who their parents are.

People with dark skin usually have ancestors (relatives who lived way back before your mom and dad) that lived in hot, sunny places (like the desert). To stay in the Sun all the time, their ancestors needed more melanin to protect themselves from the strong rays of the Sun.

Lighter-skinned people often have ancestors who lived where there was less sunlight (like the far north). Pale skin lets in more sunlight, a sure way to let skin make vitamin D.

Can You Feel That?

Melanin is not the only thing you'll find in your skin. The skin is where your body keeps its *sense of touch*. To feel things your skin has lots of nerves and special structures scientists call *touch receptors*. Here's an experiment you can do to see how sensitive your skin can be.

All you need are two pencils and a friend. The pencils should be sharp, but not too sharp. Have your friend sit down and put on a blindfold. You're ready for the experiment to begin.

◆ Holding one pencil in each hand, touch the points gently and firmly against different parts of your friend's skin. Use one or both of the pencil points at a time, but don't tell your friend how many you are using. Have her guess how many pencils you are pushing into her skin.

◆ Try the back of your friend's hand. She will feel two separate points when you press down. Then try the back of her neck. Whether you press one pencil or use both together, it will feel like one point pushing down. If you spread them farther apart, your friend will eventually feel two points. The back of the hand has more touch receptors than the neck.

Try this experiment on other parts of your friend's body. Remember to push gently with the pencils.

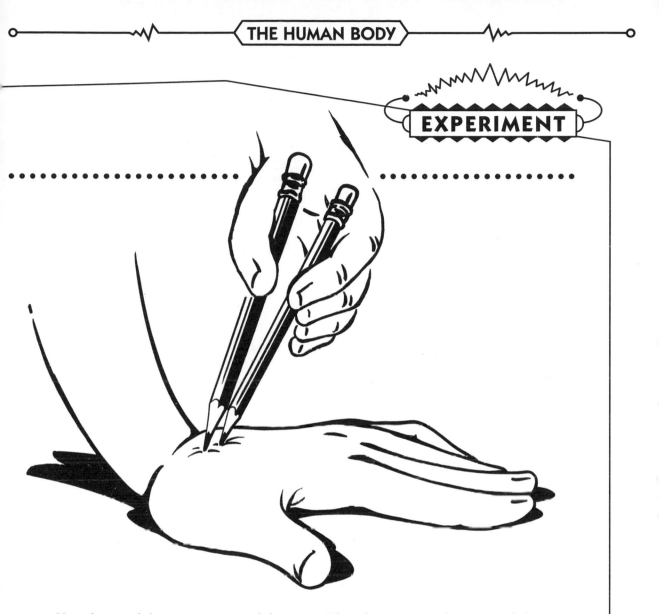

Your lips and fingers are two of the most sensitive areas of skin on your body. You may have noticed that babies often feel things with their mouths. That's because humans have lots of touch receptors on their lips.

The skin on our fingers and feet are very sensitive, too. Sensitive hands help us control our pencils. And the sensitive skin on our feet helps us keep our balance. We're sensitive where we need to be.

Eyeballs

Eye Know a Thing or Two About Science

Right now, you're reading with your eyes. If you look at this page and then look at a tree across the street, parts of your eyeballs have to change their shape.

PLEASE CONSIDER THE FOLLOWING:

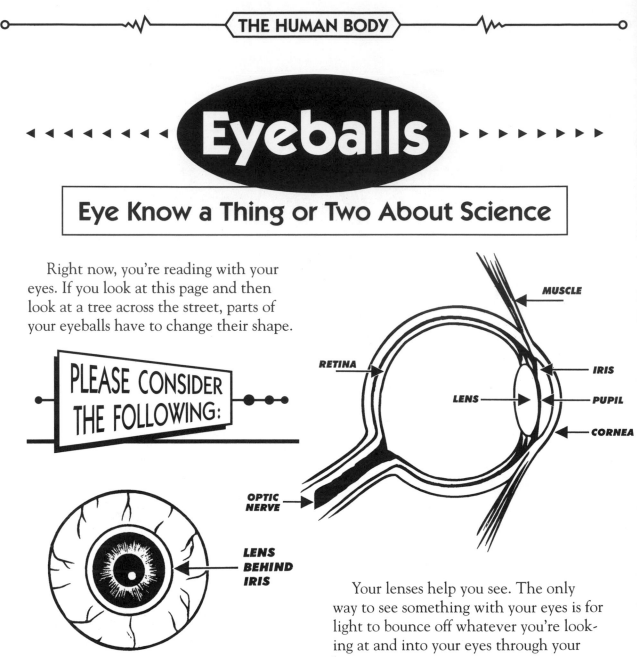

MUSCLE

RETINA

IRIS

LENS

PUPIL

CORNEA

OPTIC NERVE

LENS BEHIND IRIS

Just like a camera or a telescope, all eyes have a lens. The lenses in your eyes are clear, just like a glass lens, but they're alive. Lenses are made of living cells.

Your lenses help you see. The only way to see something with your eyes is for light to bounce off whatever you're looking at and into your eyes through your lenses.

Now, the light that goes through your lenses doesn't pass straight through. It changes direction; it bends. When light goes through a lens, it slows down. Light

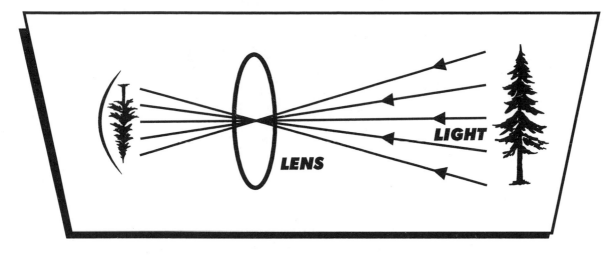

goes slower through water, glass, or plastic than through air. All lenses—whether in a camera or an eyeball—make light go a little slower. And slowing light down makes it change direction. It's like pushing a cart in the grocery store with one wheel that's sticking. It slows one side down and the cart doesn't go straight. It curves.

If your lenses bend the light just right, you will see a sharp image of whatever you're looking at. If your lenses bend too much or too little, the image will be fuzzy. When something looks fuzzy, the lenses in your eyes almost automatically change shape. They're bent by tiny muscles attached to their edges. These muscles either relax to flatten the lens or contract to squeeze it into a more curved shape.

If something is far away from us, like a sign on a building, our lenses have to curve only a little to bend light from the sign. But when you look at something close to your face, like this book, the muscles in your eye squeeze the lenses to make them curve more. That way, the light lands in just the right spot. That's what we call focusing light. Wild.

Your eyes can focus on light bouncing off this page and then focus on light bouncing off the Moon. That's not bad, considering that the Moon is 400,000 kilometers (250,000 miles) away. But to see this page, the Moon, and everything in between, the lenses in your eyes change their shape by only about a tenth of a millimeter. That's smaller than the width of the period at the end of this sentence. Just this much of a change, and you can see things right in front of you or hundreds of thousands of kilometers away. And it's all because you have muscles in your eyeballs that squeeze your lenses. Your eyes are really something, as you can see.

Build Your Own Eye

The lenses in your eyes must bend to focus the light that comes into them. But if no light enters your eyes, like when you're in a dark room, then you can't see anything. And if too much light comes in, you also can't see anything, because it's too bright and your eyes become overloaded with light. To make sure just the right amount of light enters your eyes, each eye has a muscle called the *iris*. The iris is the colored part of your eye—it can be brown, hazel, green, or blue. It opens and closes to let just the right amount of light through your lens.

To see how the iris works, you can make a model of your eye. You'll need a paper cup, wax paper, a rubber band, and a safety pin.

◆ Poke a hole in the middle of the bottom of the paper cup with the safety pin.

◆ Put a piece of wax paper over the mouth of the cup and hold it in place with the rubber band. What you now have is a model of an eye.

◆ Hold the pinhole end of the cup up to a bright lightbulb. You don't have to be very close—about a half a meter (one and a half feet) away should be fine.

◆ Now, slowly move the cup toward the lightbulb. You will see an upside-down image of the lightbulb appear on the wax paper.

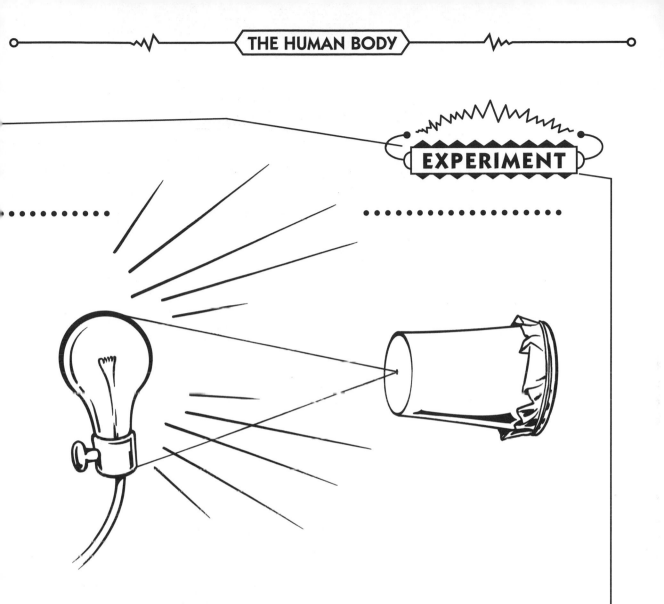

EXPERIMENT

In this model eye, the hole acts like your eye's lens. If it is just the right size, it will focus the image of the lightbulb onto the wax paper. Since you decide how big the hole should be, *you* are the iris for this model. You decide how much light gets into your "eye." To see how important the iris is, try this experiment again, making the pinhole in the cup a little bit bigger or a little bit smaller. The amount of light that gets into your "eye" makes a big difference.

Nutrition

Counting Calories the Scientific Way

Everybody is always talking about how many Calories are in their food. So what exactly *are* Calories, anyway?

PLEASE CONSIDER THE FOLLOWING:

Calories are a way to measure energy, including the energy stored in food. When the food you eat combines with the oxygen you breathe, you make energy. Scientists keep track of the energy in food by counting Calories. One Calorie in your food is equal to the amount of heat energy it takes to raise the temperature of one liter of water by one degree Celsius. Whew! That's quite a mouthful, especially for just one Calorie. But notice it's one liter, one degree, one Calorie. So it is a pretty simple way scientists have found to measure the energy in food.

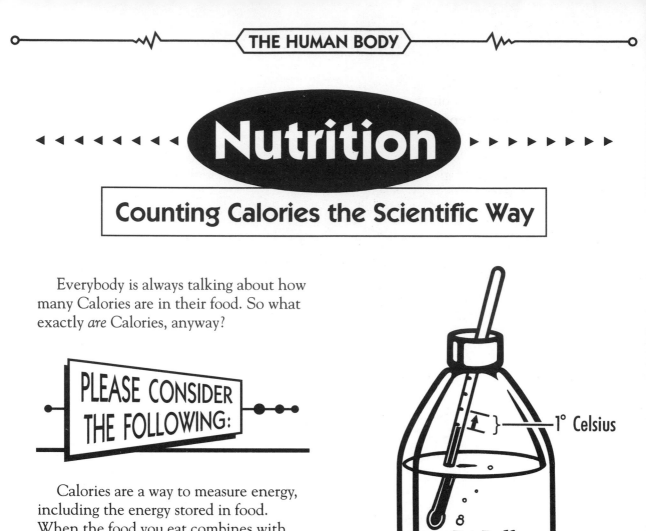

1° Celsius

Liter

When a chemical combines with oxygen, it usually gives off energy and heat. That's what happens to the chemicals in a burning candle. Candles combine wax with the oxygen in air to give off light and energy and heat. In a way, you slowly burn all the banana milkshakes and rice and beans you eat, just at a much lower temperature than a candle. Your cells take in the food and combine it with oxygen. When you burn your food, you make the energy you need to live.

So counting Calories is just a way of measuring the heat of burning food. And whether you decide to burn a hamburger in your body or in a fireplace, it will always give off exactly the same number of Calories!

Scientists measure the Calories in food by burning it in a special device called a *bomb calorimeter* (cal-oh-RIH-meh-ter). The food is sealed inside a metal container that often looks like a small paint bucket. This container is then held in a small pool of water while the food inside gets burnt to a crisp. By measuring how warm the pool gets, we can figure out how many Calories are in the food.

By the way, Calories in food are always spelled with a big C. That's the energy needed to heat a liter of water one degree. If we spell it with a small *c*, a calorie is the energy needed to heat a milliliter of water or a cubic centimeter ("cc") one degree. It's just something scientists started doing a long time ago, and it's still with us.

People my size (68 kilograms, or 150 pounds) need around 2,000 Calories a day, every day, just to get the energy to move around and keep warm. When we eat more food than we can burn, our bodies store the extra food as fat. But if we don't eat enough food, our bodies burn the fat that we've stored. Eating healthy foods is a good idea, but starving yourself isn't. You need food—and Calories—*every day* to get the energy you need for your brain and body to stay in good shape.

EXPERIMENT

Burning Nuts

The best way to see what a Calorie looks like is to burn some food. Get an adult to help you with this experiment.

◆ Take a nice hard nut, like a peanut or an almond.

◆ Unbend and rebend a paper clip into a small stand to hold your nut.

◆ Set the nut on the small bent end of the paper clip.

◆ The nut should have its long sides going up and down, like a football being held for the kicker.

Now, light a match and hold it to the upper end of the nut. It will burn just like a candle. Watch how long the nut burns. A whole peanut contains about four Calories. Remember how your nut burned the next time you eat a peanut butter and jelly sandwich. You're eating the energy of a bunch of burning nuts!

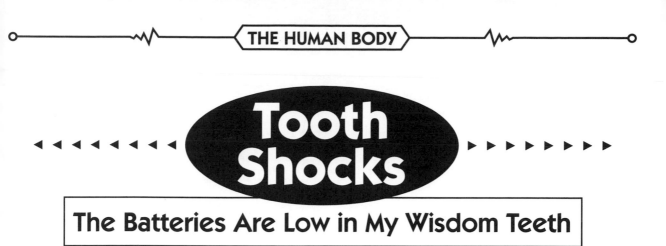

Tooth Shocks

The Batteries Are Low in My Wisdom Teeth

Have you ever bitten down on a piece of aluminum foil? If you have fillings in your teeth, it probably felt like this:

YEEEEEOUUUCHHH!

What's going on?

PLEASE CONSIDER THE FOLLOWING:

When you bite down on metal, you cause *electricity* to flow through your teeth and give your nerves a jolt. And it all has to do with the fillings in your teeth. See, your fillings are usually silver or gold. Gum wrappers and baked-potato holders are aluminum foil. All of these things are metals.

Metals, like everything else we can touch or feel in the Universe, are made up of tiny particles called *atoms*. And all atoms have even tinier particles called electrons flying around them all the time. Some atoms have electrons that stick close by, but the electrons in metal atoms aren't held on so tightly. The electrons on their atoms can jump off and check out the atom next door. They're like bees buzzing around a field of hives. They might easily fly from one hive to another. When they get to the one hive, they might move on to the next one and the next one after that.

Electricity happens when electrons move from atom to atom. One way to get electrons moving is to get two metals together in an *acid*. (An acid is a chemical that lets atoms gain or lose their electrons easily. Orange juice is an acid. So is vinegar.) If there's silver or gold in your fillings and a piece of aluminum foil in your mouth, all you need is an acid and electricity will flow.

Now, your spit—or *saliva*, as we scientists like to say—is an acid. It's not a very strong acid, but it's strong enough to get the electrons in your mouth moving and then . . .

YEEEEEOUUUCHHH!

When you bite down on foil, it's like you're making a battery in your mouth, and it electrifies the nerves in your teeth.

Conductors and Insulators

Electrons can move through the metal in your teeth, but they can't move through

just any old item. Electrical wires are not made of spaghetti.

If electrons can move through something, we call that thing a *conductor*. Metals like aluminum foil, copper pipes, gold bars, and the silver in your fillings are all conductors.

Things that aren't so good at conducting electricity are called *insulators*. We use insulators to keep electricity from going places it shouldn't—like into our bodies. Plastic, rubber, and ceramic mate-

EXPERIMENT

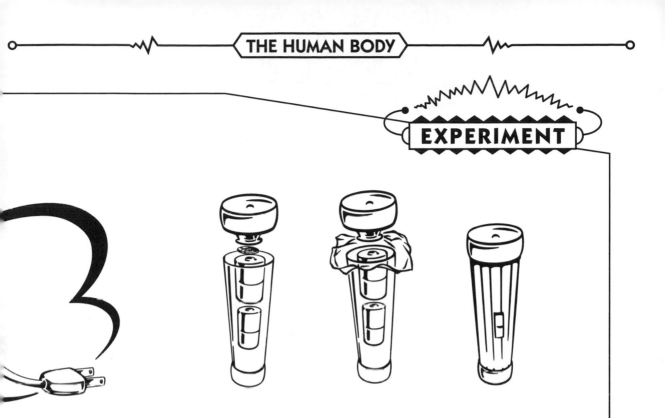

rials like plates and glasses make good insulators. That's why most electrical wires are coated in plastic or rubber.

To see how conductors and insulators work, find a small flashlight with a top that unscrews.

◆ Hold a piece of a plastic bag between the batteries and the lightbulb.

◆ Screw the top back on. Don't worry, flashlight batteries are too small to give you a shock.

◆ What happens? Does the flashlight still work? Is the plastic bag a conductor or an insulator?

◆ Now try holding a penny between the top of the battery and the bottom of the lightbulb. Does the flashlight work?

Ask yourself this: Could there be any free electrons between the bulb and the battery? Once you get the idea, lights should come on—in the flashlight, and maybe in your head.

THE LAST WORD

These are just a few of the questions we've thought about. For us scientists, there will always be more questions than answers. Every time we figure something out, or think we have, it leads us to wonder about something else. But we can't help asking these questions, trying to answer them, and thinking about them—*considering* them. Thank you for reading. And thank you for joining me in having considered the above . . . I mean, what was the following.

Absolute zero The coldest and lowest temperature there can be, at which molecules would stop moving or bouncing around; -273.2°C or -459.7°F.

Acid A large group of sour chemicals that often contain the element hydrogen and break apart easily in water. Powerful acids can cause severe burns, while a mild-mannered acid like orange juice is good to drink.

AIDS (Acquired Immune Deficiency Syndrome) A condition that people infected with HIV (Human Immunodeficiency Virus) develop, where diseases that would normally be handled by healthy immune systems make people very sick.

Archimedes A philosopher and scientist who more than two thousand years ago noticed that when he took a bath, he displaced an amount of water that took up exactly as much room as he did. When he saw that he could use displacement to understand buoyancy, Archimedes is supposed to have shouted, "Eureka!" (I've found it!)

Atoms What scientists usually think of as the smallest part of a chemical that can react with other chemicals and not be changed. They're way too small to see individually, even with a microscope.

Attraction The force that causes oppositely charged particles to draw together, such as when the north pole of one magnet and the south pole of another pull toward each other.

Aurora Glowing bands or sheets of light in the sky. As the Earth moves through space, it flies through giant clouds of tiny particles given off by the Sun. When these particles get pulled into the Earth's atmosphere, they make huge, dancing sheets of light high in the sky. They're usually seen close to the North or South Poles, where the pull of the Earth's magnetic field is stronger and the particles get pulled straight down.

Bacteria One-celled living things that can get on or in our bodies. They are sometimes harmless, but at other times they can cause disease by multiplying rapidly and attacking our cells. A single one of these is called a bacterium.

Battery A chemical source of energy with two terminals—positive and negative. A chemical reaction makes electricity flow from negative to positive when the terminals are hooked together. That's electricity.

Bernoulli's principle Named after the Swiss science guy Daniel Bernoulli, who discovered that air moving quickly creates high pressure in the direction the air is traveling, and the faster the air moves the less pressure it exerts in other directions.

Biodegradable Garbage that breaks down into soil when exposed to small organisms in the air or water. Plastics and metal are not biodegradable.

Buoyancy The tendency of things to float when they're in a fluid like water (or air). If something weighs less than an equal amount of the fluid around it, it floats. When an object is floating on the surface of a fluid, it displaces an amount of fluid that weighs exactly as much as it does.

Calorie A measurement of heat energy. A Calorie (with a capital C), which is used on food labels, is the amount of heat needed to raise a kilogram of water one degree Celsius. Many scientists prefer to use calories (with a little c)—one of these equals the amount of heat energy needed to raise just one gram of water one degree Celsius. So there are 1,000 calories in a Calorie.

Calorimeter An instrument used to measure the amount of heat given off by specific substances. This is the way scientists determine how many Calories are contained in foods.

Capillary action The movement of liquids through small spaces. They're pulled along by the stickiness of their molecules. You can see capillary action when you step into a river and the fibers in your pants get wet higher up than the level of the water.

Carbon dioxide The gas that you breathe out and plants breathe in. Animals give it off when they combine oxygen with chemicals in their food. Plants combine carbon dioxide with water and energy from the Sun to make food they can use.

Carotene A natural orange chemical in plants. It's the stuff that makes carrots orange and not purple or pink.

Cell An organized compartment of chemicals that are necessary for life. Almost all living things are composed of one or many cells.

Celsius The scientist, Anders Celsius, who based his temperature scale on water. He called the freezing point of water zero degrees and the boiling point of water one hundred degrees.

Chemicals Groups of atoms connected together are called chemicals. When two or more chemicals combine to form a new chemical, we call it a chemical reaction.

Chlorophyll The chemical found in plants that not only causes their green color but also enables them to make their own food out of sunlight, water, and carbon dioxide through a process called *photosynthesis*.

Climate How warm and wet and windy a place is over the course of the year.

Cold-blooded A way to describe animals whose body temperature goes up and down with the temperature of their surroundings.

Compass A tool that points to the Earth's magnetic North Pole. Good for finding your way in the open ocean, out of a dark forest, or just downtown.

Conductor Any material, such as a metal, that allows electric current to flow easily.

Cross section A two-dimensional view that, by cutting away a "slice" of an object, reveals what that object looks like on the inside.

Cytoplasm The jellylike fluid inside a cell.

Deciduous Trees that have leaves that are big, flat, and usually green. In moderate climates these leaves turn red, yellow, and orange, and fall to the ground every year before winter comes.

Displacement The idea, first recorded by Archimedes, that an object in water (or another fluid, like air) pushes away an amount of water equal to the size (volume) of the object.

DNA (deoxyribonucleic acid) The blueprint of life; a spiral staircase of chemicals that provides the instructions for making a living thing. There is a copy of an organism's own DNA in nearly every cell of each living thing.

Electricity The flow of electrons through something (like a wire). Electrons flow from where there are a lot of them to where they can spread out.

Electrons Negatively charged particles that buzz around atoms all the time.

Elements Chemicals that are made of just one single kind of atom.

Energy What makes things go, run, or happen. It takes energy to move the weight of a book from the floor to a table.

Enzymes Chemicals that cause specific reactions to occur when mixed with other substances. Saliva has an enzyme that helps break down the starch in food as soon as we start chewing.

Equator An imaginary line that divides a planet like the Earth in half—into northern and southern hemispheres.

Evergreen Trees with dark green needles for leaves. They lose their needles all year round instead of all of their leaves at once in the fall, like deciduous trees.

Evolution The process of natural change in the DNA of living things as they live, reproduce, and die. Each generation of a living thing is slightly different than the one before. So all species of living things slowly change the way they look and live over time.

Fahrenheit The scientist, Gabriel D. Fahrenheit, who was the first person to use mercury in a thermometer. He based his temperature scale on a mixture of the salt ammonium chloride and called the coldest temperature he could reach zero degrees.

Friction The resistance to movement of two objects touching each other. It makes your hands heat up when you rub them together.

Germ Any tiny living thing that makes another, different living thing get sick. Also, the first form of a living thing—wheat germ is like a seed of wheat.

Gills The part of a fish that takes in the oxygen a fish needs from the water and releases carbon dioxide. Gills are like fish lungs.

Golgi body The part of a cell that gets rid of waste by packaging up cell garbage into little cellular bags. They are very important to cells. (Imagine what living in your house would be like if you never got rid of the trash!)

Gravity The force that pulls all things toward each other.

HIV (Human Immunodeficiency Virus) It attacks human white blood cells and leads to AIDS.

Hormones Chemicals your body makes that control everything from how fast your body grows to how come you feel happy, sad, angry, or scared.

Inertia The tendency for things that are still to stay still and for things that are in motion to continue moving until something—or someone—slows them down.

Insulator Any material, such as plastic or rubber, that doesn't permit the flow of electricity and is often used to coat electrical wires.

Iris The colored part of the eye; made up of muscles that open and close the pupil (the black circle in your eye) to control how much light gets inside.

Landfill A special kind of garbage dump where the garbage is buried between layers of soil, often covering wetlands or other low-lying areas.

Lysosome The part of the cell responsible for taking apart things that are too big or complex for the cell to use.

Magnet A piece of rock or metal that can push or pull on another piece of metal without touching it.

Mammals Warm-blooded, furry, usually non-egg-laying creatures that feed their babies milk.

Matter What things that we can touch and feel are made of. Ice and clouds are made of matter; light is not.

Melanin The tiny particles of color in human skin that absorb the Sun's harmful radiation—protecting the body and making skin dark or tan. Every person has different amounts of melanin, which means some people have a lot (and have darker skin), while others have a little (and have lighter skin).

Membranes Thin, flexible layers that surround cells and keep them separate from one another.

Mitochondria The energy factories of a cell.

Molecules When two or more atoms connect together, they form molecules.

Nerve A group of special cells that makes up the nervous system, the body's communication network. Nerve cells transmit electrical impulses to and from the brain and body.

Nucleus A structure inside animal and plant cells that contains most of the DNA.

Operculum The part of a fish that covers and protects its gills. There is one on each side.

Oxygen The gas we need to breathe in to live and that plants need to give off to live. The atmosphere contains about 21% oxygen.

Plankton The tiny organisms that live in the top (sunny) part of the ocean. They can be zooplankton (like animals) or phytoplankton (like plants). Everything in the ocean depends on plankton to survive.

Protein A chemical in food that your body needs to grow and repair itself. Meat, eggs, milk, and rice and beans are all good sources of protein.

Reptiles Cold-blooded, scaly animals that usually lay shelled eggs. Alligators are reptiles; dinosaurs were reptiles.

Repulsion The tendency of like particles to push away from one another, such as when the north poles of two magnets are placed close together. All electrons push away from each other, because they are all negative.

Ribosome The protein factory of a cell. The DNA passes the protein blueprints that a cell needs to its ribosomes, where proteins get made.

Saliva The watery stuff in your mouth. It contains chemical enzymes that break down starches into sugars.

Supernova An exploding star that has run out of fuel. It collapses in on itself, causing a huge explosion that eventually creates new stars and planets.

Touch receptors The endings of the neurons of the nervous system that are located in the skin and transmit the sensations of touch and pain to the brain and body.

Vacuum A space that has nothing in it. There can be partial vacuums—spaces with nearly nothing in them. Carpet-cleaning appliances and drinking straws use partial vacuums.

Virus The cause of many diseases, including the common cold and AIDS. They are tiny groups of chemicals that get inside cells and multiply like crazy. It's hard to say if viruses are truly living things or not.

Warm-blooded A way to describe animals that keep their body temperature nearly constant.

Wetland Any place where the land is soaking wet all or part of the year.

Xanthophyll A yellow-colored chemical in plants that helps convert the Sun's energy into chemical energy the plant can use.

Weights and Measures of Science

LENGTHS OR DISTANCES

Meters

If you're a kid, a meter is about the distance between your hands when they're stretched all the way out. It's a little longer than a yard (39.37 inches). A soccer field is 100 meters. An American football field is 100 yards. Soccer fields are a little longer. A kilometer is 1,000 meters or 10 soccer fields. A kilometer is 0.6214 miles—that's a little more than half a mile, a little less than three-quarters.

WEIGHTS AND MASSES

Kilograms

If you weigh a liter of water, it's a kilogram. Sometimes we just say a "kilo" (KEE-loh). So the liquid in a two-liter soft drink bottle would weigh almost exactly two kilograms. A kilo is a little more than 2 pounds, right around 2.2 pounds. A kilogram is a thousand grams. A mosquito weighs about a gram. A metric ton is just 1,000 kilograms. That's easy. That would be right around 2,200 pounds, or 200 pounds more than an old English ton.

One more thing about weight. Things have weight only when there's gravity around, like here on Earth. But things always have mass, even in space. It's a part of nature you may not have ever thought about. Big boulders and spacecraft are hard to push around, even when they're floating around in space, because they always have mass. So, the kilogram is a unit of mass. Pounds originally started out as units of weight, although some scientists use a unit called *pound-mass*. It has to be written *lbm*. I'll tell you as a scientist, the difference between pound and pound-mass has caused many, many problems.

VOLUME

Liters

A liter is the volume of 10 by 10 by 10 centimeters. That's a cube with each side a tenth of a meter long. And a liter of water weighs 1 kilogram. Pretty cool, huh?

TEMPERATURE

Celsius

A system of measuring temperature, with 0 degrees being the freezing point of water and 100 degrees being the boiling point. To convert Celsius to Fahrenheit, multiply by 9/5 and add 32.

Fahrenheit

A system of measuring temperature, with 32 degrees being the freezing point of water and 212 degrees being the boiling point. To convert Fahrenheit to Celsius, subtract 32 and multiply by 5/9.